Remarkable Natural Material Surfaces and Their Engineering Potential

Michelle Lee
Editor

Remarkable Natural Material Surfaces and Their Engineering Potential

 Springer

Editor
Michelle Lee
Mechanical Engineering
McCormick School of Engineering
  at Northwestern University
Evanston, IL, USA

ISBN 978-3-319-34570-3   ISBN 978-3-319-03125-5 (eBook)
DOI 10.1007/978-3-319-03125-5
Springer Cham Heidelberg New York Dordrecht London

Printed on acid-free paper

Springer is part of Springer Science+Business Media (www.springer.com)

Foreword

It is my great pleasure to introduce to you this book that contains a series of articles about the surfaces of several species in our natural world.

We are luckily living on the only blue globe in the universe so far understood by modern astronomy, sharing the same blue sky and water with other species also

Dikilitaş, in *Sultanahmet Meydanı*, Istanbul, Turkey, an example showing the inspiration from nature on human society civilization

nursed by planet Earth. At the same time that we enjoy our bread and butter, we admire the awesome power of nature that has created the rich, beautiful, and complicated yet mysterious botanic, marine and animal systems.

Thousands of years of human civilization has only explored small corners of the mighty natural world formed through millions of years of evolution. Nature demonstrates its laws peacefully and gracefully sometimes, but loudly at other times. Ancient Greeks showed their admiration of nature by creating gods related to natural powers. The ancient Egyptian language is among the oldest languages; it uses many symbols from nature. A good example is *Dikilitaş*, or the Obelisk of Theodosius, which was the Ancient Egyptian obelisk of Pharaoh Tutmose III. It was re-erected by Roman emperor Theodosius I in the fourth century AD in *Meydanı* or *Sultanahmet Meydanı*, the Hippodrome of Constantinople in Istanbul. The Obelisk silently explains the close link between early civilization and the study of birds, plants, and animals, from which ancient wisdom tirelessly tells stories of past glories.

Nature has nursed human culture and knitted unforgettable stories. At a school in ancient China, handsome Liang Shanbo met Zhu Yingtai. He thought Yingtai was a bright young fellow like him, and they became the best of friends. However, Yingtai was a beautiful girl disguised as a boy in order to attend the school, because girls were denied school education during that time. Eventually Shanbo realized Yingtai was a girl and wanted to marry her, but the Zhu family had already promised her to another family by matchmaking despite her unwillingness. Shanbo soon passed away due to extreme desperation. On the way to her marriage ceremony, Yingtai prayed at the grave of Shanbo, and the tomb magically cracked open. Yingtai jumped in, and suddenly, a pair of colorful butterflies flew out into the blue sky to continue their never-dying love story.

Wright Brothers National Memorial, Kill Devil Hills, North Carolina, the world's first controlled powered flight. Courtesy of US Department of the Interior, National Park Service

Nature-inspired technological development dates back to the days of the domestication of plants, grass, insects, birds and animals in order for people to have food and basic living supplies. Food production demanded continuous improvement of tools; travel and the exchange of goods required the innovation of faster and faster vehicles; the desire for victory in wars here and there necessitated high-speed missiles and long-distance information communication, just to name a few. Nature ignites people's imagination and spurs the invention of machines and systems that have never existed to meet these needs. Airplanes, submarines, and trains, as well as countless technological innovations like these, emerge from the images of their natural counterparts.

People dreamed of flying like a bird; some even sacrificed their lives when jumping into the wind with a pair of wings. It was not until the Wright brothers demonstrated the world's first controlled powered flight in 1903 in Kill Devil Hills, North Carolina, USA, that the aviation age was effectively launched. From then on, airplanes and spacecrafts have brought people's dreams to record-breaking heights.

Every object, natural or man-made, has its interior material structure and surface, and many biological and physical functions are accomplished through surfaces. The surface of one material system forms an interface with that of another, such as the interface between water and its container, pen and paper, shoes and the ground, blood and the blood vessel, meshing gear teeth, the tire and road, everywhere and every system where components are in contact and relative motion. In many cases, surfaces and interfaces determine system life, efficiency, functionality, reliability, and security. For example, frozen rain can result in heavy ice shells on power-line surfaces, leading to power transmission system damage. A typical ice-rain storm in Northeast United States may result in damages of about $4–6 billion dollars. Surface fouling can retard productivity in agriculture and cause inefficiency of ship navigation. Friction at interfaces and wear of surfaces result in energy waste and equipment damage. The significance of understanding surfaces can never be overestimated.

Nature has inspired the creation of countless products by means of mimicking natural systems; it is also unfolding its secrets of surfaces and stimulating future surface-based technologies to help people win the challenges of energy crises, environmental issues, and system reliability problems. Many are waiting to be done.

ME 346, Introduction to Tribology, is a course teaching the phenomena, theories, and principles related to surfaces and their interactions at Northwestern University, Evanston, Illinois, USA. Tribology is the science of interaction of surfaces in contact and relative motion, where understanding the science of surfaces is one of its major objectives. It is open to both juniors and seniors and graduate students. Twenty-three students enrolled in this class in the fall quarter of 2012. While teaching principles of tribology, such as contact, lubrication, friction and wear, I encouraged students to look around and see what nature tells us about surfaces and how people are developing novel technologies based on the understanding gained from the studies of natural surfaces. Nearly one half of the class was involved in learning the inspiration that natural surfaces could offer as the subject of their class projects. The class projects of these students resulted in 13 interesting articles, ranging from plants to animals and insects. The modification of these articles led to this book,

What micro and nano mysteries are behind the beauty of Jiu Zhai Gou Valley, China?

edited by Ms. Michelle Lee, one of the 346 students. Five students participated in the work, and others allowed Michelle to re-write the articles based on their original project reports. After ME 346, Michelle spent about 10 months of her free time to organize issues related to the book, write her own chapters, edit articles from other students, and communicate with the publisher. This book could not have been done without her effort. It is worth mentioning that Ms. Caroline E. Hartel, another ME 346 student, assisted Michelle in developing the proposal of this book's writing.

As presented in this book, lotus leaves, rice leaves, gecko pads, butterflies, diatoms, etc., have the microscopic wonders behind their pretty appearances. Many more astonishing surfaces of plants, algae, bacteria, animals, insects, stones, etc., are out there waiting for young students to explore, study, and dream up inventions from. Enjoy reading and thinking!

Evanston, IL, USA   Q. Jane Wang, Ph.D.

Preface

There is an epidemic that has swept the entire nation—one that most are oblivious to. It is the epidemic of overly fast-paced, technology driven lives. Now, more than ever, Americans lead lives of busyness and constantly being on the run, with cell phones, email, and tablets at the ready.

At some point we have to wonder… Could we be eliminating something valuable from our lives with this kind of lifestyle? But, of course, just as soon as these thoughts form in our heads, they disappear at once as the phone chirps, and we grab it to see who texted us.

The truth is that there *are* many valuable things we ignore every day. In fact, there are a thousand, a billion, an *infinite* amount of things, perhaps right outside our windows or in the ponds near our houses. They may even be found amidst the mass of forest trees along the highway we took today on our commute to school or work, or in the park that we walked our dogs in. With such a rapid increase in the rate of technology usage, it is easy to forget nature, the master inventor.

But we always have the ability to stop, and truly look.

> When we stare this deeply into nature's eyes, it takes our breath away, and in a good way, it bursts our bubble. We realize that all our inventions have already appeared in nature in a more elegant form and at a lot less cost to the planet. Our most clever architectural struts and beams are already featured in lily pads and bamboo stems. Our central heating and air-conditioning are bested by the termite tower's steady 86 degrees F. Our most stealthy radar is hard of hearing compared to the bat's multifrequency transmission. And our new 'smart materials' can't hold a candle to the dolphin's skin or the butterfly's proboscis. Even the wheel, which we always took to be a uniquely human creation, has been found in the tiny rotary motor that propels the flagellum of the world's most ancient bacteria.
>
> –Janine M. Benyus, *Biomimicry: Innovation Inspired by Nature*

Nature is a bountiful source of inspiration and intelligence, and many have been using it for scientific exploration for ages. Such examples include gecko pads, lotus leaves, butterfly wings, sharks, and more. This book presents 13 different kinds of natural surfaces—many of which we can easily find near our houses—and their respective science and engineering values. The chapters in this book were written by students in Professor Q. Jane Wang's Introduction to Tribology class taught at Northwestern University in Evanston, IL. USA. All passionate about nature, the contributing student authors strive to share their understanding with the communities around them.

After reading the fascinating examples of natural surfaces in the following pages, may you be inspired to slow down, get some fresh air, and discover for yourself all that nature has to offer.

Evanston, IL   Michelle Lee

Acknowledgements

I would like to thank Professor Q. Jane Wang for her mentorship from beginning to end, as well as her inspiring teaching throughout the Introduction to Tribology course at Northwestern University.

I would also like to take this opportunity to express sincere gratitude to the participants of this book:

Mindie Chu, Ignacio Estrada, Shiqi Luohong, and Yunho Yang, for their hard work and contribution that made this book possible;

Caroline Hartel, for her enthusiasm and her help with getting this book off the ground, as well as allowing me to use her draft as a basis for the chapter on lotus leaves;

And Elizabeth Bifano, Rose Gruenhagen, Joseph Park, and Christopher Timpone, for allowing me to consult their drafts in writing the chapters on gecko pads, spider silk, snake skin, and dragonfly wings, respectively.

Contents

1 Blood Clots and Vascular Networks: Self-Healing Materials 1
  Michelle Lee

2 Shark Skin: Taking a Bite Out of Bacteria 15
  Michelle Lee

3 Mother-of-Pearl: An Engineering Gem 29
  Michelle Lee

4 Diatoms: Glass Ornaments of the Earth's Waters 41
  Mindie Chu

5 Lotus Leaves: Humble Beauties 53
  Michelle Lee

6 Dragonfly Wings: Special Structures for Aerial Acrobatics 65
  Michelle Lee

7 Moth Eyes: A New Vision for Light-Harnessing Efficiency 79
  Michelle Lee

8 Botanical Leaves: Groovy Terrain 91
  Ignacio Estrada

9 Snake Skin: Small Scales with a Large Scale Impact 103
  Michelle Lee

10 Gecko Pads: A Force to Be Reckoned with 115
   Michelle Lee

11 Butterfly Wings: Nature's Fluttering Kaleidoscope 127
   Shiqi Luohong

12   Frog Skin: A Giant Leap for Engineering Applications 135
Yunho Yang

13   Spider Silk: A Sticky Situation.. 145
Michelle Lee

Index.. 159

Michelle Lee

The Gift of Regeneration and Renewal

Though many childhood memories fade over the years, there are some that remain vivid as we enter into adulthood. Though these memories may range from bitter to sweet and differ per person, most can recall that *one* accident, be it falling off the bike and getting scraped from head to toe by the pavement, breaking or fracturing a limb during a rowdy game of street hockey with some neighbors, or getting a huge cut doing something just plain dumb. This is most likely the story that gets shared at every family reunion or as an interesting personal anecdote. Whatever the story, talking about it in adulthood is usually painless—often even humorous and sweet. We are able to share our stories nostalgically, because what is left of the accident is a mere scar—many times not even that. This is a direct result of our bodies' phenomenal self-healing and renewing capabilities.

On the other hand, unlike us, most of the objects we use on a day-to-day basis are on a one-way road to deterioration. In other words, the scratch we made on our car the other day from a bad parking job will be there in a year, along with all the other dents and scratches we make in between. The web-like crack in our window from the neighbor's rogue baseball will be there to obstruct the nice view of our backyard for weeks to come until we replace it. Objects must be replaced, and replaced, and replaced.

Engineers hope that one day, this will not be the case for certain applications in which self-repairing ability would be particularly advantageous. This area of research is called self-healing materials, and the following sections will give an overview of the bio-inspiration behind self-healing materials as well as current and future applications.

M. Lee (✉)
Mechanical Engineering, McCormick School of Engineering
at Northwestern University, Evanston, IL 60208, USA
e-mail: MichelleLee2013@u.northwestern.edu

M. Lee (ed.), *Remarkable Natural Material Surfaces and Their Engineering Potential*,
DOI 10.1007/978-3-319-03125-5_1, © Springer International Publishing Switzerland 2014

Self-Healing Materials

What Does It Mean to Be Self-Healing?

A material is self-healing if it can recover from damage without external intervention. Other words that have been used interchangeably with self-healing include autonomic-healing, autonomic-repairing, and self-repairing (Ghosh 2009). Perhaps one of the biggest differences between most living organisms and engineered materials is the ability to adapt and heal in response to damage and degradation. Engineered materials generally lack the inherent ability to fix themselves, and they deteriorate over time due to degradation, whether it is in the form of fatigue, creep, brittle fracture, wear, and so on. This deterioration is irreversible, leading many materials to catastrophic failure (Nosonovsky 2012).

Current Engineered Materials

According to van der Zwaag et al., engineered materials' inability to self-heal is a result of its underlying *damage prevention* paradigm, as opposed to nature's underlying *damage management* paradigm. The *damage prevention* paradigm, which dominates the development of all engineering materials, is defined by a design strategy of most effectively delaying the onset of deterioration and damage. As a result, today's engineering materials will only experience an increase in damage or, in the best case, remain at a constant level.

On the other hand, nature's inherent *damage management* paradigm allows damage to be unproblematic, since the damage can be self-healed (van der Zwaag et al. 2009). The lack of autonomous self-healing mechanisms in engineering materials necessitates human intervention and efforts to remedy effects of deterioration, which most often take the form of welding, resin injection, and applying reinforcement patches. These methods are typically imperfect and leave the repair site as the weakest point. Therefore, materials that can self-heal at the micro- and even nano-scale level are of great interest to scientific communities (Guimard et al. 2012).

Advantages of Self-Healing Capabilities

Healing mechanisms of living organisms are extremely complex, making it difficult to mimic them. In addition, copying the healing mechanisms of nature is unwise, because the intrinsic character of engineering materials has to be taken into account when designing self-healing materials (van der Zwaag et al. 2009). However, nature is a source of bio-inspiration for researchers who are seeking to create self-healing materials for the future of engineering (Nosonovsky 2012). Such self-healing materials are "smart" materials: materials designed to have a predetermined and controlled response to a particular stimulus. Theoretically, using self-healing materials would fail less often, have a longer lifetime, and decrease the frequency of maintenance as well as the costs associated with such tasks (Guimard et al. 2012). Furthermore, self-healing materials would ensure repair of damages that occur in remote or hidden locations (Ghosh 2009).

'Bleeding' Approach to Self-Healing

Blood Clotting as Bioinspiration

A simple observation of nature is that animals self-heal by means of a 'bleeding' mechanism (Trask et al. 2007). In humans, this is often encountered at a young age at the first cut or wound. From then on, bleeding simply becomes a part of life, the unavoidable and rather unpleasant intermediate step between getting hurt and healing. However, bleeding leads to blood clotting, a process critical to the body's healing ability, which involves interactions between the damaged tissue, blood plasma, and blood platelets. Blood clotting was observed with much fascination by early scientists and philosophers who marveled at the body's natural ability to heal wounds through the clotting mechanism. Galen (129–199 AD) was one of them, describing threads that he observed in blood clotting. Malpighi also noticed threads and nerve-like networks in 1666 (Ferguson et al. 2010).

Today, we know much more about the details of mammalian blood clotting. The threads that were observed and marveled at by Galen and then Malpighi are known as fibrin fibers, which are composed of protein and produced when a wound breaches the endothelial cell lining of blood vessels. This breach initiates a sort of cascading series of active enzymatic reactions that involve inactive precursors called clotting factors, ultimately resulting in fibrin (see Fig. 1.1). Blood clotting is remarkable in that it rarely malfunctions. In fact, the body continues to take measures to prevent further damage, rapidly removing activated enzymes once fibrin is produced and avoiding clots in healthy blood vessels by breaking fibrin down in any undamaged areas. Overall, the blood clotting process involves around 80 coupled biochemical reactions of platelet cells and enzymes (Trask et al. 2007). Clotting is thought to be a self-regulating procedure, and this self-regulation is elemental to the success of its two basic features. One feature is its threshold response; clotting will only occur in a blood vessel that is substantially damaged. The other feature is its locality, occurring only in a confined region in the immediate vicinity of the damage (Runyon et al. 2004).

Despite the wealth of knowledge about the specific mechanisms involved in blood clotting that had not been available in Galen and Malpighi's time, the fascination with the process still continues today, as blood clotting has become a source of bioinspiration for the development of self-healing materials. Using blood clotting as a biomimetic platform would be difficult considering the number of reactions involved and its complexity (Runyon et al. 2004), as evident in the previous section. However, it has inspired the use of liquid healing agents that leak from fractured openings of damaged hollow fibers or capsules as a form of autonomic self-healing (Trask et al. 2007).

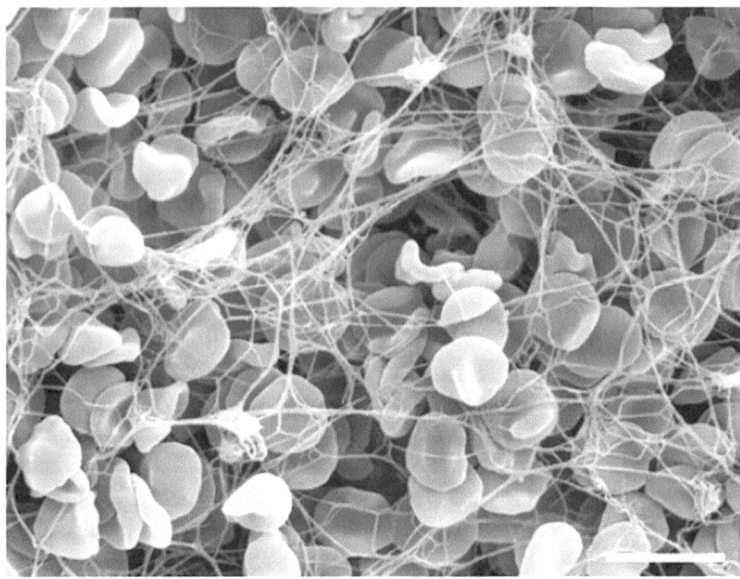

Fig. 1.1 Electron microscope image shows erythrocytes braided by fibers of fibrin in a blood clot. Scale bar: 10 μm (Reprinted from Ferguson et al. (2010), Fig. 1. With kind permission from Springer Science + Business Media)

Encapsulation

The capsule strategy features a microencapsulated healing agent dispersed throughout a structural composite matrix. Within the matrix is a catalyst that can polymerize the healing agent upon contact. With these two components in place, the self-healing process proceeds as such: microcapsules of the healing agent rupture as a crack approaches, the healing agent flows into the crack plane via capillary action, the healing agent comes into contact with the catalytic chemical trigger, causing polymerization to occur, which finally seals the crack faces (White et al. 2001). Figure 1.2 depicts the capsule strategy.

According to Dr. Scott White, a professor in aerospace engineering at the University of Illinois, and his team—the first researchers to demonstrate the success of this self-healing concept in 2001—there were numerous aspects that had to be considered while designing such a process. Such factors included the strength of the interface between the matrix and microcapsule, as well as the stiffness and toughness of the microcapsules themselves. Another design parameter was the thickness of the capsule walls. For example, the walls would break during processing if the walls are too thin, but thick walls would not rupture at the critical moment when the crack approaches. Using micromechanical modeling, the researchers examined the interaction between a microcapsule and a crack, which was very complex and three-dimensional. These analytical studies combined with optical and scanning electron

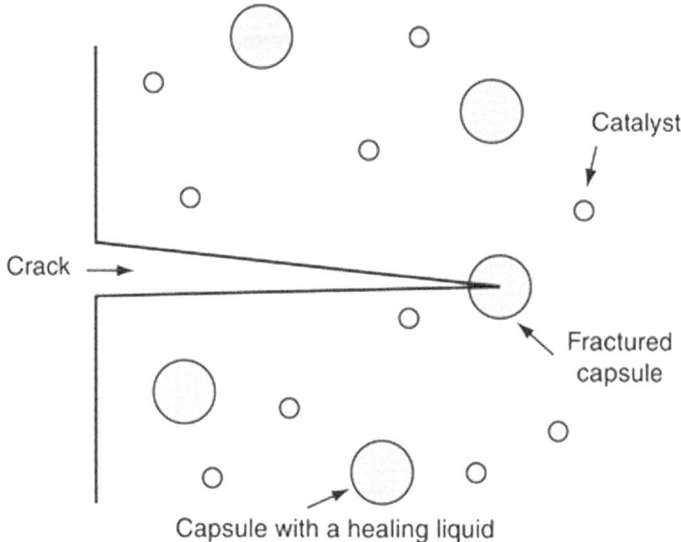

Fig. 1.2 Embedded microcapsules and catalysts enable self-healing of cracks (Reprinted from Nosonovsky (2012), Fig. 1. With kind permission from Springer Science+Business Media)

microscopy confirmed the success of their self-healing mechanism. They were able to obtain time sequences of images that show a rupturing microcapsule and healing agent flowing into the crack plane (White et al. 2001).

Another important component of the self-healing concept was the chemistry behind the polymerization of the healing agent. After identifying the key requirements needed in the reaction—low monomer viscosity, low volatility, long shelf life, and low shrinkage upon polymerization, among others—White et al. chose to use ring-opening metathesis polymerization (ROMP). In ROMP, a transition metal catalyst, or Grubbs' catalyst, is used alongside dicyclopentadiene (DCPD) microencapsulated in sizes ranging from 50 to 200 μm within urea-formaldehyde shells to prevent premature polymerization while the composite is being produced. The resulting reaction was then observed by environmental scanning electron microscopy (ESEM) and infrared spectroscopy, which revealed evidence proving that polymerization of the healing agent was in fact initiated by the transition metal catalyst (White et al. 2001).

There are several advantages to this particular strategy. One is that the self-healing mechanism provides local repair; the autonomic control is site-specific. Another advantage is that multiple healing events can occur, because the polymerization catalysts are living, meaning that they have unterminated chain-ends. Finally, sealing microcracks through this mechanism would prevent stress corrosion cracking, moisture swelling, and other forms of deterioration encouraged by the environment (White et al. 2001).

Compared to control samples without the self-healing mechanism embedded, the self-healing composite performed well in testing. The three control samples were epoxies with either no catalyst and no microcapsules, no catalyst with the presence of microcapsules, or no microcapsules with the presence of the catalyst. Results of a load–displacement curve showed the self-healing composite to experience 75 % recovery of the original fracture load, while all control samples could not continue to carry any more load. Despite some limitations, such as catalyst stability in certain environments and crack-healing kinetics, White et al. anticipated self-healing composites to be a potentially powerful solution to the limited lifetimes of polymeric structural materials (White et al. 2001). Indeed, since White et al.'s demonstration of their microencapsulation concept, rapid advancement has followed in this approach (Sottos et al. 2007).

Hollow Glass Fibers

While the use of hollow glass fibres for self-healing had been explored in various materials such as bulk concrete and bulk polymers, during the same year as White et al.'s progress with their microencapsulation approach, Bleay et al. pioneered the use of hollow glass fibres (HGF) in polymeric composites. Though different in structure compared to microencapsulation, HGF was also inspired by the body's bleeding mechanism—this time, specifically the arteries (Trask et al. 2007). Bleay et al.'s work, though found to display a better combination of mechanical and storage qualities than that of microencapsulation, was limited in its own way: the amount of healing agent available (Pang and Bond 2005).

Pang and Bond, who specifically aimed to achieve higher internal volumes than its predecessors (Pang and Bond 2005), created a composite material featuring the embedment of hollow fibres containing healing agents that, when tested, displayed a flexural strength restoration of 97 %. Similar to microencapsulation, when a crack fractures a fibre, healing agent or resin is released and reacts with a hardener to induce polymerization in the crack, sealing it (Ghosh 2009). In addition, highly apparent substances such as UV fluorescent dye are released upon HGF fracture, allowing damage needing attention to be identified more quickly. The HGF produced by Pang and Bond can be seen in Fig. 1.3, and a damaged area made visible by UV dye can be seen in Fig. 1.4.

HGF is not only advantageous in its capacity to hold healing agents, but it also offers structural reinforcement to composite materials (Pang and Bond 2005). Using tailored HGF in composite matrices have been shown via analytical and finite element modeling to have gains surpassing 200 % in transverse strength and specific rigidity. Furthermore, HGF-embedded composites absorb energy better than systems with solid fibres due to the crushing of HGF, leading to a higher tolerance to impact damage (Hucker et al. 1999).

The theory behind the gains in specific flexural rigidity and strength of HGF compared to solid fibres is based on the work of Burgman, Watson, and Watson and

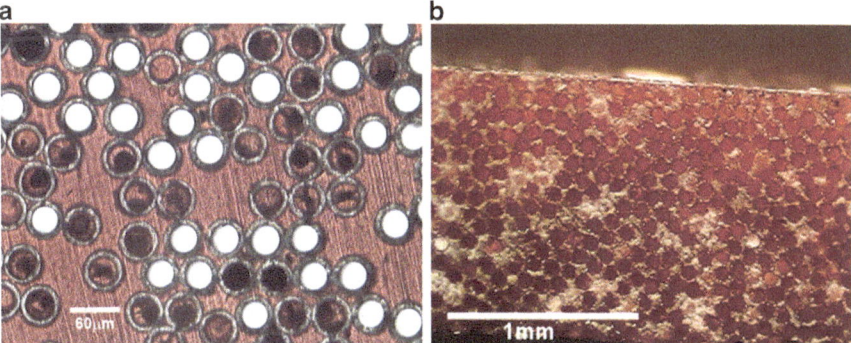

Fig. 1.3 Optical micrographs by Pang and Bond of hollow glass fibres (HGF) and composites produced at Bristol: (**a**) HGF with external diameter measuring 60 μm and 50 % hollowness, and (**b**) the HGF featured in (a) embedded in an epoxy matrix (Hexcel 913) (Reprinted from Pang and Bond (2005). With permission from Elsevier)

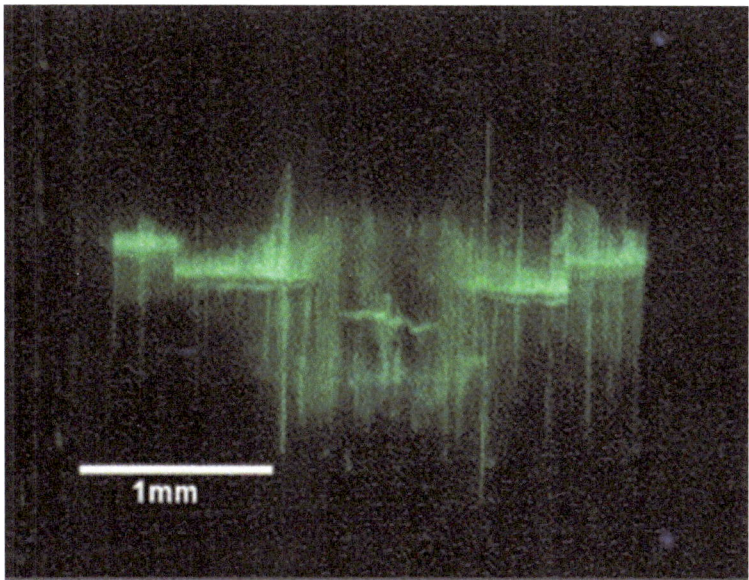

Fig. 1.4 Ultra-violet mapping technique (UVMT) illuminates the UV fluorescent dye (Ardrox 985) that bled from the fibres upon impact. UV dye effectively highlights damaged regions, making them easier to detect and evaluate (Reprinted from Pang and Bond (2005). With permission from Elsevier)

Farrow, where the flexural rigidity of a composite beam that contains hollow fibres is as follows:

$$R_h = \frac{E_s^* b t_s^3}{12\left(1-K^2\right)^2}$$

where b is the width of the composite beam, t is the thickness, and E_s^* is the effective modulus of the laminate containing solid fibres.

The fibre hollowness ratio is:

$$K^2 = d^2/D^2$$

where d is the inner diameter of the hollow fibre, and D is the outer diameter.

Furthermore, the relative rigidity ratio (h indicative of hollowness and s indicative of solidness) is given by the following:

$$\frac{R_h}{R_s} = \frac{1}{\left(1-K^2\right)^2}$$

It can be noted here that since $K^2 \leq 1$, $(1-K^2) \leq 1$ and the rigidity and critical load for buckling will increase for HGF compared to solid fibres at identical weights of glass.

Furthermore, the relative maximum bending stress ratio and ultimate strength ratio are derived to be the following, respectively:

$$\frac{\sigma_{max\,h}}{\sigma_{max\,s}} = \left(1-K^2\right)^2$$

$$\frac{\sigma_{ult\_h}}{\sigma_{ult\_s}} = \left(1-K^2\right)$$

Calculations based on the equations above at varying hollowness ratios reveal HGF-embedded composites to be up to 100 times more rigid and capable of taking 10 times the amount of bending moment before the material fails (at hollowness ratio K^2 of 0.9) (Hucker et al. 1999).

Some disadvantages of this approach include the necessity of fibers to be fractured in order for the healing agent to be released and the need for more than one step in fabrication. There are also many advantages, however, such as the availability of higher volumes of healing agent, the ease with which fibres can be mixed and tailored with the already existing, conventional reinforcing fibres, and the variation allowed in resin choice or activation methods (Ghosh 2009). The reinforcement that hollow glass fibres themselves contribute to composite materials, as shown in detail through the theoretical work of multiple researchers, is also a great advantage. Though their design is not intended to eliminate permanent damage within a composite, it can prevent further propagation (Pang and Bond 2005).

Microvascular Network

Despite the work of researchers such as Pang and Bond, who created more volume in the microfibres of their composite material for healing agents, the problem yet remains: what happens when all microcapsule or microfibre volumes are ruptured, and the supply of healing agent is depleted? This scenario highlights the limited reality of the 'bleeding' mechanism. Once again, though, nature provides yet another source of inspiration for the solution of such a daunting problem: vascular systems (Andersson et al. 2008).

Vascular systems are ubiquitous in nature, from the dendritic design of leaf veins shown in Fig. 1.5 to the system probably most widely known by humans—the circulatory system. While vascular systems in plants are responsible for the transport of water and photosynthates (Ueno et al. 2006), blood, oxygen, and nutrients, among other components, are transported in humans. As the primary barrier to the outer environment, keeping the skin intact is a very important task, requiring any damage to be quickly repaired. Through the capillary network of the circulatory system, necessary components for healing are carried to the damaged site, and a clot forms. As we know, the skin is capable of healing more than once; it would be quite the catastrophe if it couldn't! This ability can be attributed to the vascular nature of the circulatory system (Toohey et al. 2007). Vascular systems in biological organisms are complex and do many things including, but not limited to, enabling self-healing, distributing fuel, and regulating internal temperature, but at the most basic level,

Fig. 1.5 Veins of a leaf

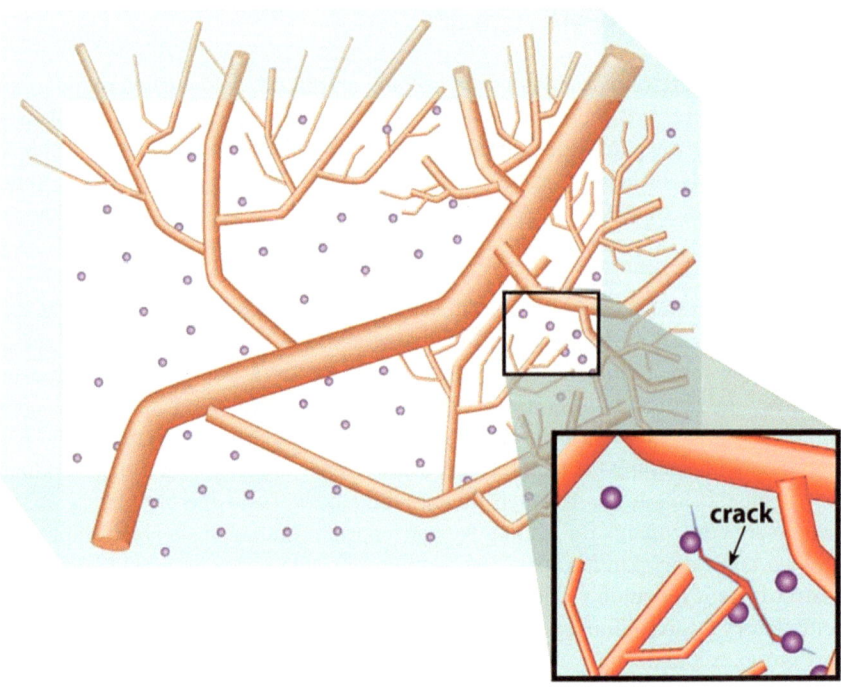

Fig. 1.6 Conceptual image of embedded microvascular network that would supply healing agent (Reprinted from Toohey et al. (2009), Fig. 1. With kind permission from Springer Science + Business Media)

they all transport fluid from a reservoir to more remote locations. Vessel size and branching of these networks have naturally evolved to distribute and maintain the transported fluid using the minimum amount of power necessary (Trask et al. 2007).

The difference between microcapsule or microfibre concepts and circulatory concepts is apparent in their structures, which can be compared to compartments and networks, respectively (Andersson et al. 2008). While the compartmentalized approach lacks continuity and replenishment of healing agents, the network approach, shown in Fig. 1.6, offers a renewable, mobile supply to regions of damage (Trask et al. 2007). Through an access point, healing agent can be restocked to avoid depletion (Hume 2013).

Engineering Applications

Corrosion Protection

The engineering applications for which self-healing materials can be useful are vast, because extending the life of materials is in the interest of and could benefit many industries. Innovations in materials technologies have contributed much to the

Fig. 1.7 Corrosion on the bow of an oil tanker (Photo by Haus)

enhancement of the reliability, cost effectiveness, and performance of engineering structures, and the development of autonomic self-healing materials could be key to further improvement in the performance of such engineering structures (Trask et al. 2007).

In 2005, previously mentioned Dr. Scott White founded Autonomic Materials™. Autonomic Materials™ is a company whose technologies are based on Dr. White's work with self-healing materials at the world-famous Beckman Institute (Autonomic Materials™ 2013). According to the company's chief executive officer, initial prospective applications for the self-healing technology featuring microencapsulation are associated with corrosion protection. Specifically, self-healing coatings would be applied to marine assets like docks, ships, and oil/gas platforms, structures that are frequently damaged through collisions or during transportation. For instance, a self-healing coating that had been applied to the ship in Fig. 1.7 could potentially have protected it from the corrosion it experienced. Given that estimated costs associated with corrosion reach up to $500 billion worldwide every year, Autonomic Material™'s self-healing coatings would help reduce unnecessary costs, not to mention the headache that the industry experiences in having to tend to rigs and other structures, which are usually in remote locations (Hume 2013).

Potential Applications for Microvascular Network Concept

Dr. Ian Bond and his team have been hopeful in one day applying their vascular approach to self-healing materials to solve problems and meet demands that are otherwise unmet by currently available materials. As the head of the aerospace engineering department at the University of Bristol in the UK, Bond predicts it will still be a while before their self-healing composites will be used in the construction of airplanes, due to the industry's high testing standards. However, he has other plans for the vascular self-healing mechanism that don't involve a flying machine.

Self-repairing glass is one example, and Bond and his team have succeeded in demonstrating success. Some applications for this technology include car and military vehicle windows, which would provide clear vision for drivers getting away from sites of conflict. Additionally, Bond estimates that wind turbines may feature the self-healing material in 5–10 years. Mobile phone screens are another example, although long-lasting materials for this application do not seem necessary since trend and style of the phone usually dictate how long it is used (Hume 2013).

The technology is not yet perfect; setbacks range from chemical instability to unreliability in extreme conditions, in addition to ones that have been pointed out in previous sections of this chapter. However, each advance in self-healing materials is one step closer to achieving materials capable of full recovery (Hume 2013).

Through the dedication and hard work of researchers and innovators, self-healing materials are becoming more and more of a reality. Originally inspired by blood clots and vascular networks, self-healing materials can revolutionize the field of engineering and technology with a variety of applications ranging from self-repairing glass, corrosion protection, innovative aerospace composites, and more.

References

Andersson HM, Keller MW, Moore JS, Sottos NR, White SR (2008) Self healing polymers and composites. In: van der Zwaag S (ed) Self healing materials. An alternative approach to 20 centuries of materials science. Springer, Netherlands, pp 19–44. doi:10.1007/978-1-4020-6250-6_2

Autonomic Materials (2013) World-recognized technologies. http://www.autonomicmaterials.com/our_company/. Accessed 15 Aug 2013

Autonomous Materials Systems: Beckman Institute for Advanced Science and Technology (2007) Microvascular systems. http://autonomic.beckman.illinois.edu/mvac.html. Accessed 14 Aug 2013

Ferguson J, Nürnberger S, Redl H (2010) Fibrin: the very first biomimetic glue—still a great tool. In: von Byern J, Grunwald I (eds) Biological adhesive systems. Springer, Vienna, pp 225–236. doi:10.1007/978-3-7091-0286-2_15

Ghosh SK (2009) Self-healing materials: fundamentals, design strategies, and applications. In: Ghosh SK (ed) Self-healing materials: fundamentals, design strategies, and applications. Wiley-VCH Verlag GmbH & Co. KGaA, Weinheim, pp 1–28. doi:10.1002/9783527625376.ch1

Guimard NK, Oehlenschlaeger KK, Zhou J, Hilf S, Schmidt FG, Barner-Kowollik C (2012) Current trends in the field of self-healing materials. Macromol Chem Phys 213(2):131–143. doi:10.1002/macp.201100442

Hucker MJ, Bond IP, Foreman A, Hudd J (1999) Optimisation of hollow glass fibres and their composites. Adv Compos Lett 8(4):181–189

Hume T (2013) Heal thyself: the 'bio-inspired' materials that self-repair. http://edition.cnn.com/2013/02/22/tech/self-healing-materials. Accessed 15 Aug 2013

Nosonovsky M (2012) Self-repairing materials. In: Bhushan B (ed) Encyclopedia of nanotechnology. Springer, Netherlands, pp 2382–2385. doi:10.1007/978-90-481-9751-4_217

Pang JWC, Bond IP (2005) A hollow fibre reinforced polymer composite encompassing self-healing and enhanced damage visibility. Compos Sci Technol 65(11–12):1791–1799

Runyon MK, Johnson-Kerner BL, Ismagilov RF (2004) Minimal functional model of hemostasis in a biomimetic microfluidic system. Angew Chem Int Ed 43(12):1531–1536. doi:10.1002/anie.200353428

Sottos N, White S, Bond I (2007) Introduction: self-healing polymers and composites. J R Soc Interface 4(13):347–348. doi:10.1098/rsif.2006.0205

Toohey KS, Sottos NR, Lewis JA, Moore JS, White SR (2007) Self-healing materials with microvascular networks. Nat Mater 6:581–585. doi:10.1038/nmat1934

Toohey KS, Sottos NR, White SR (2009) Characterization of microvascular based self-healing coatings. Exp Mech 49(5):707–717. doi:10.1007/s11340-008-9176-7

Trask RS, Williams HR, Bond IP (2007) Self-healing polymer composites: mimicking nature to enhance performance. Bioinspir Biomim 2(1):1–12. doi:10.1088/1748-3182/2/1/P01

Ueno O, Kawano Y, Wakayama M, Takeda T (2006) Leaf vascular systems in C3 and C4 grasses: a two-dimensional analysis. Ann Bot 97(4):611–621. doi:10.1093/aob/mcl010

Van der Zwaag S, van Dijk NH, Jonkers HM, Mookhoek SD, Sloof WG (2009) Self-healing behaviour in man-made engineering materials: bioinspired but taking into account their intrinsic character. Philos Trans R Soc A 367(1894):1689–1704. doi:10.1098/rsta.2009.0020

White SR, Sottos NR, Geubelle PH, Moore JS, Kessler MR, Sriram SR, Brown EN, Viswanathan S (2001) Autonomic healing of polymer composites. Nature 409:794–797. doi:10.1038/35057232

Shark Skin: Taking a Bite Out of Bacteria

Michelle Lee

A Change in Reputation

Danger, jaws, life-threatening, and *fear* are probably a few words that come to mind when you think of sharks like the one in Fig. 2.1. These words are often associated with sharks, and it makes sense given that almost a fifth of Americans are more terrified of sharks than any other animal (Public Policy Polling 2013). After all, if their gigantic bodies are not enough to make someone shudder, one look at their razor-sharp teeth and cavernous mouths definitely would. In addition to the menacing appearance of sharks, this popular view is encouraged in part by many movies like *Shark Night* and *Jaws*, which are famous for their chilling dramatizations of sharks. Most recently, a documentary was aired by Discovery Channel called *Megalodon: The Monster Shark Lives*, which detailed the possible return of Submarine, a 67-foot-long shark that terrorized the seas during the Miocene era. Though the documentary received much backlash from viewers and specialists alike who insist that the claims of the film are fake, such documentaries and popular movies like the ones mentioned above unmistakably affect the general view of sharks (National Geographic 2013).

Though the danger of certain sharks is very real—numerous shark attacks in which victims have been maimed or killed have indeed been documented over the course of history—not all match the aggressive, bloodthirsty images portrayed in theatres. For example, the basking shark pictured in Fig. 2.1 is actually a passive filter-feeding shark, which means that it consumes food such as fish eggs, larvae, and small crustaceans by straining water through its gill slits. They are not considered to be dangerous to observers, being quite tolerant of boats and divers (Florida Museum of Natural History 2013).

M. Lee (✉)
Mechanical Engineering, McCormick School of Engineering
at Northwestern University, Evanston, IL 60208, USA
e-mail: MichelleLee2013@u.northwestern.edu

M. Lee (ed.), *Remarkable Natural Material Surfaces and Their Engineering Potential,*   15
DOI 10.1007/978-3-319-03125-5_2, © Springer International Publishing Switzerland 2014

Fig. 2.1 Basking shark (*Cetorhinus maximus*) (Photo by Chris Gotschalk)

Furthermore, sharks have even been receiving attention from the scientific community as of late for their contribution to the health and well being of humans. Specifically, shark skin has been found to possess remarkable features that have important applications in the medical and engineering arenas. This chapter will explain how and why these 350 million year old, speedy swimmers have experienced a turnaround in reputation due to their skin (Friedmann et al. 2010). Perhaps by the end of this chapter, your mind will change about them, too.

Basics of Shark Skin

A Force and Displacement Transmitting Tendon

Shark skin is an active skin, meaning it is very closely connected to the animal's thrust producing muscles that enable its undulatory, or wave-like, movement (Naresh et al. 1997). When sharks swim at a slow speed, it does not bend its body as much as it does when it swims faster. Wainwright et al. found that when swimming slowly, the lemon shark (*Negaprion brevirostris*) had a radius of curvature of 38 cm, decreasing to a radius of curvature of 20 cm when traveling fast. Because its muscles are attached to the bone on one side and skin on the other, shark skin is much like its own external tendon, shortening and lengthening based on curvature and transmitting forces and displacement to the tail (Wainwright et al. 1978).

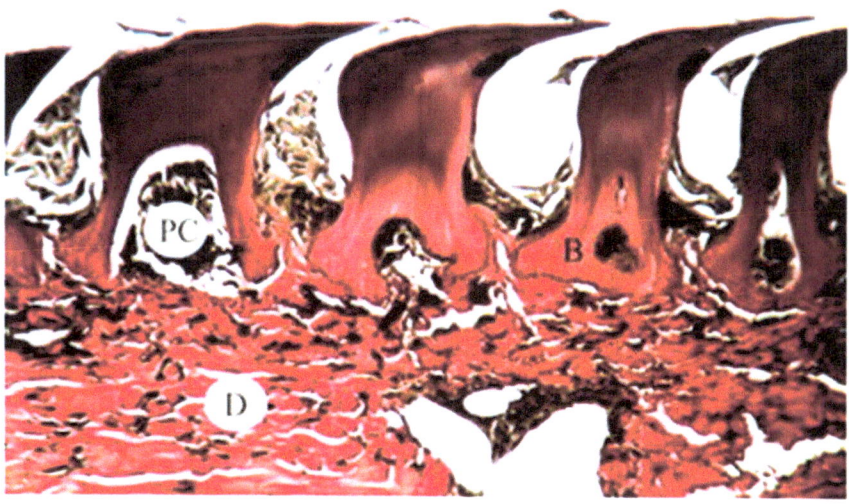

Fig. 2.2 Lateral view of the skin of a shortfin mako shark (*Isurus oxyrinchus*), showing the pulp cavity (PC), the dermis (*D*) in which the placoid scales are anchored, and the base (*B*) of the scales (Reprinted from Lang et al. (2012), Fig. 1. With kind permission from Springer Science + Business Media)

Skin Structure

This important organ is of a complex structure, a composite that is collagen fiber reinforced and pliant. Like any other vertebrate skin, it is composed of flesh, then dermis, and finally the outermost layer, which is called the epidermis. The dermis has two layers: the upper stratum laxum and the lower stratum compactum (Naresh et al. 1997). The stratum compactum consists of layered, helically wound collagen fibers, forming an exoskeleton around the whole shark body (Lang et al. 2008).

Covering the whole shark are pointy placoid scales commonly referred to as dermal denticles (Gilbert 1984). The reason why these placoid scales are called dermal denticles is because they have a dentine vascular core and an outer enamel layer, much like human teeth (Magin et al. 2010). The dentine core is made of apatite, a hard, crystalline mineral, giving denticles the strength of steel and hardness of granite (ReefQuest Centre for Shark Research 2003). Dermal denticles are regularly spaced at intervals of about 100–300 mm with heights ranging from 200 to 500 mm (Singh et al. 2012). While the base of each denticle is embedded in the stratum compactum, the end of each scale, known is the crown, is exposed to water (Lang et al. 2008). The crowns interlock at the exposed surface, and each crown has a set of riblets. The number, dimensions, and arrangement of riblets depend on the species; for example, the blacktip shark (*Carcharhinus limbatus*) has five riblets spaced 0.065 mm apart with a height measuring 0.029 mm, while the shortfin mako (*Isurus oxyrinchus*) has three riblets spaced 0.041 mm apart with a height measuring 0.012 mm (Lang et al. 2012). Figure 2.2 shows a lateral view of the skin of a shortfin mako shark.

Dermal denticles have a range of purposes. They reduce mechanical abrasion, reduce drag while swimming, and protect sharks from ectoparasites and predators. Interestingly, dermal denticles have also been found to aid in food processing. Through observational and experimental methods, Southall and Sims found that *Scyliorhinus canicula*, commonly known as dogfish, utilized its skin to reduce food size. Dogfish and other sharks have been known to engage in certain behaviors while feeding, such as sucking and spitting food out of their mouths and rapidly shaking their heads. Southall and Sims were the first to recognize that dogfish placed food so that it snagged on their denticles to anchor the prey as they jerked their head away to break pieces off (Southall and Sims 2003). Another intriguing attribute of shark skin denticles that has recently been the focus of many studies is their ability to prevent biofilms, and this topic will be explored throughout the remainder of the chapter.

Biofilm Prevention

Sessile vs. Planktonic

Denticles serve several purposes for sharks, and a very important one is bacteria prevention. Bacteria in natural environments can be found in two states: planktonic and sessile. While planktonic or free-floating bacteria promote the spread of bacteria to new places, sessile bacteria is conditioned to withstand adverse environmental situations. Sessile bacteria attach to surfaces and form a biofilm—a structured community of bacteria (Glinel et al. 2012).

The formation of a biofilm in an aquatic environment can be broken down into three general phases: the induction, growth, and stationary periods. The induction phase is characterized by the adsorption of inanimate organic matter to a surface, which forms a film layer with negative charge and high adhesion. The growth period occurs in two stages called reversible and irreversible adhesion. During the reversible adhesion stage, bacteria seem to be attached to the surface initially, but they can move freely and be washed away easily. At this point in time, bacteria adheres to the surface using forces from water flow, interfacial tension, gravity, electrostatic interaction, and van der Waals forces. However, during the irreversible adhesion stage of the growth period, bacteria secrete extracellular polymeric substances (EPS) to adhere permanently to the surface (Bai et al. 2013). EPS is known less formally as slime, and it is made of various components such as carbohydrates, uronic acids, proteins, and deoxyribonucleic acid (Chapman et al. 2014). Then, during the stationary period, spores, larvae, and propagules of fouling organisms adhere to the surface, eventually evolving into a complex biofilm 2 or 3 weeks later (Bai et al. 2013).

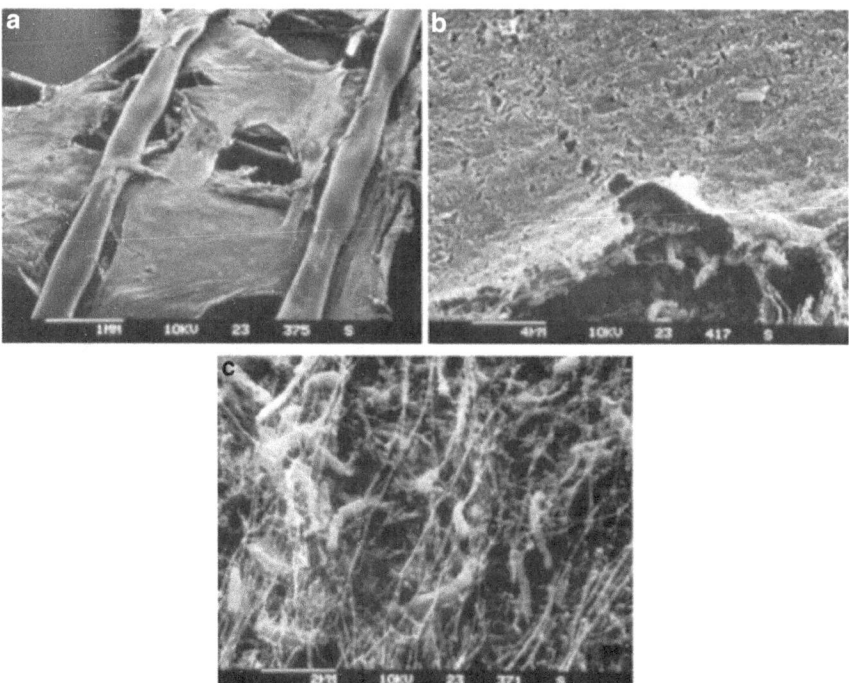

Fig. 2.3 Scanning electron micrographs of a biofouled reverse osmosis membrane show: (**a**) a spacer in the membrane completely covered in biofouling, (**b**) magnification revealing a sheet of extracellular polymeric substances (EPS) with bacteria visible just underneath, and (**c**) further magnification of the layer of bacteria, which are embedded in a fibrillar tangle that protects them from biocides (Reprinted from Flemming (2002), Fig. 1. With kind permission from Springer Science + Business Media)

The Dangers of Biofilms

Biofilms can be found on many surfaces, from underwater vessels to ultrafiltration systems (Chung et al. 2007). Biofilms are also ubiquitous in industrial settings as well, developing on a wide variety of surfaces such as pipes and membranes. In nuclear power plants, biofilms generate blockages and corrosion in condenser cooling tubes. In the water supply industry, pipes are laden with biofilms and can eventually form fungi, populations of mussels, and dangerous pathogens. Water purification membranes such as the one shown in Fig. 2.3 are susceptible to increased hydraulic resistance and decreased efficiency due to the formation of biofilms (Bixler and Bhushan 2012).

They are particularly detrimental in the biomedical arena, as they are known to cause many infections, including osteomyelitis, cystic fibrosis pneumonia, native valve endocarditis, and more (Glinel et al. 2012). Infections from *Staphylococcus*

Fig. 2.4 Scanning electron micrograph of a catheter infected with a biofilm of *Candida albicans* and *C. glabrata* (×5,000) (Reprinted from Maki and Tambyah 2001, Fig. 2)

aureus are also commonly developed from heart valves, sutures, and cochlear implants (Harris and Richards 2004). Biofilms pose the largest threat to people with implanted materials such as prosthetic heart valves, orthopaedic devices, and urinary catheters (Stewart and Costerton 2001). Figure 2.4 shows a scanning electron micrograph of a dense biofilm on the surface of a urinary catheter. Because they are often mechanically, physically, and chemically constructed to encourage the biogrowth of tissue, they simultaneously promote the growth of biofilms, which adhere to the host tissues and/or implants (Chung et al. 2007). Biofilms are also found on common hospital surfaces like doors, surgical tools, beds, and countertops, threatening not only patients, but healthcare professionals as well (Magin et al. 2010).

Compared to their planktonic counterparts, sessile communities are formidably resistant to antibiotics, hydrodynamic shear forces, and biocides—about 1,000 times more resistant (Glinel et al. 2012). One reason why antibiotics are unable to eliminate biofilms is that they cannot penetrate the entire thickness of the biofilm. Sessile communities also elicit immune responses from hosts by releasing antigens, but the antibodies are unable to destroy the biofilm (Costerton et al. 1999). Though antibiotics may kill planktonic bacteria released by the sessile community, it will not rid the host of the biofilm itself, which can then act as a nidus, or natural reservoir, for the next infection. Because of their ability to endure such hostile environments, biofilms typically persist until they are surgically removed from the host (Stewart and Costerton 2001).

Fig. 2.5 Environmental scanning electron microscope (ESEM) image shows the details of denticles on the surface of the bonnethead shark (*Sphyrna tiburo*). Structure shows common ridges running longitudinally. Scale bar: 50 μm (Reproduced with kind permission from Oeffner and Lauder (2012), Fig. 4)

Scientists have looked to methods of biofilm prevention as opposed to treatment for several reasons. First, attempts to completely eradicate biofilms with the use of antibiotics have resulted in the development of antibiotic-resistant bacteria, otherwise known as superbugs (Chung et al. 2007). Furthermore, disinfectants and antibiotics can cause severe environmental damage and toxicity (Glinel et al. 2012).

How Dermal Denticles Prevent Biofilms

As an aquatic animal, sharks are perpetually exposed to bacteria, algae, and other forms of contamination from marine organisms (Bhushan 2011). However, their dermal denticles—"little skin teeth" ribbed longitudinally with grooves in the direction parallel to local flow of water as mentioned earlier, such as in Fig. 2.5—prevent fouling organisms from adhering to the shark's skin. The rough texture that denticles impart to shark skin does this in several ways: by reducing the amount of surface area available for bacterial organisms to attach, creating an unstable surface that repels microbes, and forming a moving target for organisms as it constantly flexes in response to variations in internal and external pressure (Biomimicry Institute 2007).

Also, the spacing between scales traps air, forming air pockets that repel biofouling organisms (Malshe et al. 2013).

Dermal denticles also accelerate water flow at the surface, decreasing the contact time of organisms and limiting their ability to settle (Biomimicry Institute 2007). This occurs due to the drag-reducing abilities of dermal denticles (Bixler and Bhushan 2012). Fluid drag can be examined in two different forms: pressure and friction drag. Pressure drag is what one experiences when walking through water and is the reason why it often feels difficult to do so. This type of drag is caused by the necessity to move water out from in front of the body to the sides and back in order for forward movement to occur (Bhushan 2012). As a shark swims, pressure drag is dependent on the attachment or separation of flow. Ideally, flow separation should not occur, because it generates a pressure difference in front of and behind the shark's body, inducing greater drag. One of the primary ways in which sharks, among many more marine animals, prevent high pressure drag is the shape of their bodies, which is very smooth, tapered, and streamlined (Lang et al. 2012). The other form of drag, friction drag, is associated with the molecular interactions within the fluid and between the fluid and a surface. Friction drag is dependent on multiple things, including velocity and viscosity of fluid. In turbulent flow that is fully developed, flow is highly random as vortices eject and twist into each other in the cross flow direction (Bhushan 2012). Famously known as the shark skin effect, dermal denticles decrease overall drag in turbulent flow by dampening the cross flow velocity generated by ejected vortices, which is one of the two components of turbulent flow—the other one being main stream velocity (Friedmann et al. 2010).

Engineering Applications

Sharklet AF™

Sharklet AF™ is the first micropatterned texture designed to prevent bacteria from colonizing and migrating (Reddy et al. 2011). Designed by University of Florida researcher Anthony Brennan, the Sharklet AF™ pattern was inspired by the diamond-like patterns he found on shark skin. Shark skin was relatively clear of organisms compared to other animals such as whales, and Brennan initially hoped to reduce the buildup of algae on submarines for use on Naval marine vessels (Sharklet™ 2010). Its effectiveness has motivated Brennan to apply the bacteria-inhibiting pattern to healthcare products, other commercial products, and boats (Pogoreic 2012).

Clinical Applications of Sharklet AF™

One study performed by Chung et al. demonstrated the effectiveness of the Sharklet AF™ technology in inhibiting *Staphylococcus aureus* on Silastic® brand silicone elastomer PDMSe, a biomaterial used in a multitude of devices such as pacemaker leads, tubing, and catheters. Researchers chose *S. aureus* because of the match they found between its size and the dimensions of the Sharklet-patterned surface (3 μm height with 2 μm width and spacing); they hypothesized that bacteria with dimensions ranging from ~1 to 2 μm would not be deterred by the texture. *S. aureus* was also chosen because of its frequent association with nosocomial, or hospital-acquired, infections (Chung et al. 2007).

Comparing both smooth and Sharklet AF™ PDMSe surfaces that were subject to the growth of *S. aureus* over the course of 21 days, researchers found that the smooth PDMSe samples were significantly more covered in bacteria than the textured ones. The progression of bacteria coverage can be seen in Fig. 2.6, where the area of coverage on the smooth surface was about 40 % higher than that of the Sharklet pattern.

Chung et al. believe the bacteria-inhibiting effect to be a result of the physical obstacles that the topographical surface presents to bacteria. With such protruding structures in the way, small clusters of bacteria growing in the recesses of the texture are discouraged from expanding into larger colonies. It was only until day 21 that the bacteria were found to be growing in layers—a tactic employed to overcome the protrusions and connect to other isolated clusters. Smooth surfaces do not offer such deterrence, explaining why biofilm development started to be seen on day 7 (Chung et al. 2007). Furthermore, according to Sharklet Technologies, the protruded surfaces of Sharklet AF™ require bacteria to expend too much energy to signal to other bacteria and colonize, inhibiting their expansion. Bacteria intend to establish large colonies and, when they are unable to do so, seek out other surfaces on which to colonize or simply die (Sharklet™ 2010).

In another study, Reddy et al. examined the effect of Sharklet's micropattern on the inhibition of *Escherichia coli*, particularly for reducing catheter-associated urinary tract infections (CAUTI). With around 30 million urinary catheters inserted into patients each year and approximately 95 % of urinary tract infections a result of urinary catheter usage (Brooks et al. 2013), the prevention of infection without the use of harmful and toxic antibiotics would increase the quality of life for many patients. In this study, Reddy et al. tested the effect of variations of the Sharklet pattern on the colonization and migration of *Escherichia coli*, the most common uropathogen, across silicone elastomer, the material most used for urinary catheters. Three variations of the Sharklet micropattern were used: protruding Sharklet 2×2 (width and spacing 2 μm each), recessed Sharklet 2×2, and protruding Sharklet 10×2 (width of 10 μm and spacing of 2 μm). The fourth surface was the control: smooth, with no texture (Reddy et al. 2011).

After growing *E. coli* on the four surfaces and performing a rinse procedure to remove free-floating bacteria, results showed a significant variation in the colonization of the uropathogen between the textured and smooth surfaces. The three samples with the Sharklet pattern displayed statistically significant reductions in the percentage of area covered by the bacteria. The compiled data for percent reduction was about 47 % in bacteria colony number (colony-forming units, or CFUs), 45 % in area coverage, almost 80 % in colony size, and over 81 % in migration. These impressive results support the possibility of using Sharklet as a promising method of biofilm inhibition (Reddy et al. 2011).

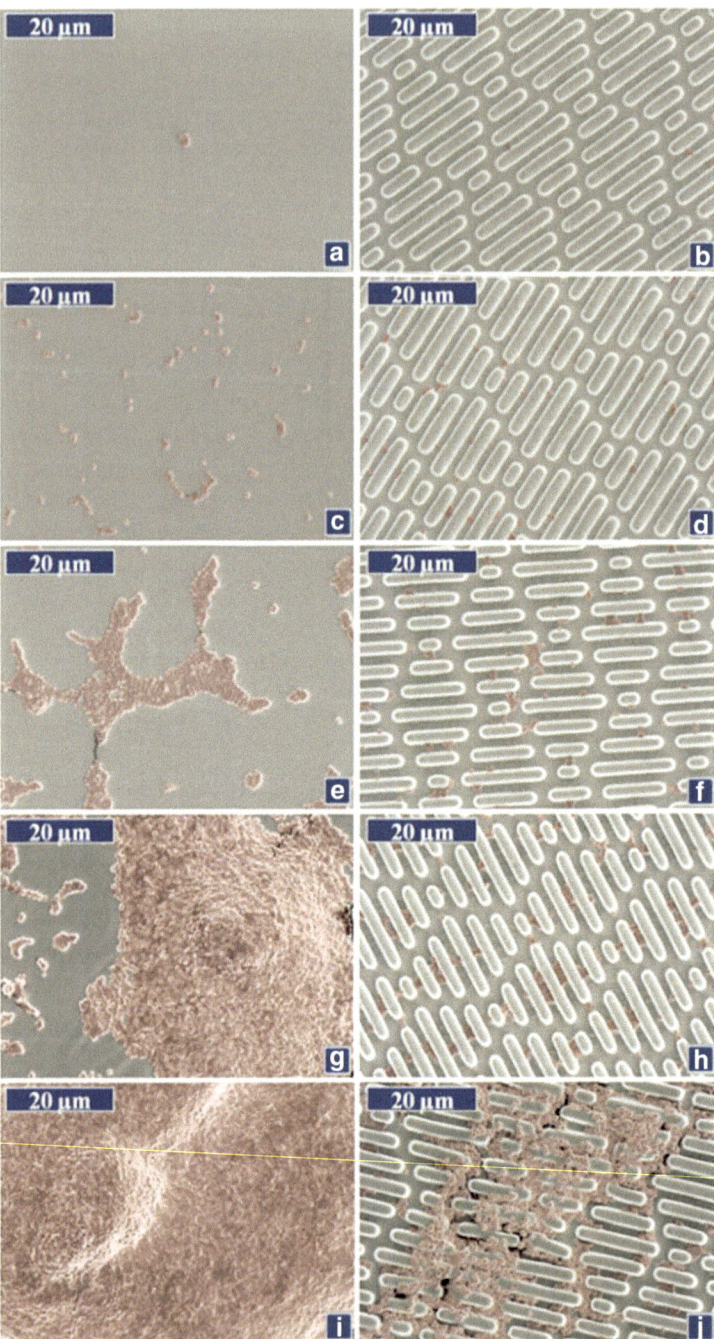

Fig. 2.6 Scanning electron micrograph images show the progression of *S. aureus* over 21 days on PDMSe surfaces (*left*, smooth; *right*, Sharklet AF™): (**a, b**) 0 days, (**c, d**) after 2 days, (**e, f**) after 7 days, (**g, h**) after 14 days, and (**i, j**) after 21 days. The Sharklet AF™ PDMSe surface shows much less progression of *S. aureus* compared to the smooth surface (Reprinted from Chung et al. (2007), Fig. 5. With kind permission from Springer Science + Business Media)

Though further research is necessary in order to study the capabilities of Sharklet under *in vivo* conditions, the strategy of using micropatterned textures like Sharklet to prevent biofilm infections from *S. aureus*, *E. coli*, and other bacteria may truly power the paradigm shift away from toxic, superbug-developing antibiotics towards effective micropatterned textures. *In vivo* success has the capability of enhancing the lives of many patients and reducing the occurrence of nosocomial infections, which are contracted at a rate of 1.7 million times per year in the United States alone (Pogoreic 2012). If a surface topography such as Sharklet could prevent *Staphylococcus aureus*, *Escherichia coli*, and other deadly bacteria, for the first time ever, sharks may truly become lifesavers.

Other Engineering Applications

Marine vessels have been plagued with issues of biofouling for thousands of years, dating back to the ancient Phoenicians, who were the first recorded inventors of anti-fouling lead coatings. Seventeenth century sailors used copper-containing metals to prevent biofouling, and in the twentieth century, antifouling paints consisting of biocides became the primary method for resisting biofouling organisms.

One such biocide that became a popular ingredient in antifouling paints was tributyltin, or TBT. Although TBT-based paints were effective, they were eventually banned because of the hazard they presented to the environment and non-target organisms (Magin et al. 2010). The effect of TBT on marine environments has been found in organisms from bacteria to fish and other mammals. For example, shell abnormalities in oysters were discovered to have been caused by exposures to TBT that were as low as 0.002 µg/L. During the time that TBT was commonly used, populations of gastropods were decreasing in areas of high marine vessel traffic. Discussion of the effect of TBT on humans was initiated when the biocide began to be found in human blood and liver, indicating that TBT was contaminating seafood designated for human consumption (Dafforn et al. 2011). The damage that biocides and metals such as lead inflict on the environment has inspired the search for environmentally friendly but effective antifouling coatings (Magin et al. 2010).

One reason why a successful antifouling agent is desired is because of the losses that ships and marine organizations incur due to biofouling. According to the International Maritime Organization (IMO), biofouling organisms like algae, barnacles, and slime that are attracted to the dark, submerged surfaces of boats increase drag resistance and decrease speed, racking up 40 % more in extra fuel costs (Lee 2009). In fact, annual costs associated with biofouling problems are estimated at $1 billion (Magin et al. 2010).

In the face of this large environmental and economical problem, applying the shark skin texture to surfaces of boats has proved advantageous and effective. Using this strategy has reduced the fouling of boat surfaces by 67 % compared to non-textured surfaces. A reduction in the amount of organisms adhered to a boat's surface has many positive effects aside from being cleaner. It reduces the amount of invasive species transported from one location to another and does not require environmentally toxic chemicals for cleaning. Boats textured in the shark pattern are also more energy efficient. Applying Sharklet AF™ could eliminate the unnecessary costs incurred due to fuel loss, since boats would be free of drag-inducing organisms (Biomimicry Institute 2007).

The number of biomedical and engineering applications in which mimicking shark skin would be beneficial is truly staggering. Not only is it capable of reducing fuel costs and protecting marine wildlife from hazardous chemicals, shark skin technology can protect humans from dangerous nosocomial infections and potentially save lives. Shark skin is a perfect example of the bounty of knowledge and inspiration we can receive from nature if we look closely.

References

Bai XQ, Xie GT, Fan H, Peng ZX, Yuan CQ, Yan XP (2013) Study on biomimetic preparation of shell surface microstructure for ship antifouling. Wear 306(1–2):285–295, Elsevier B.V., Amsterdam

Bhushan B (2011) Biomimetics inspired surfaces for drag reduction and oleophobicity/philicity. Beilstein J Nanotechnol 2:66–84. doi:10.3762/bjnano.2.9

Bhushan B (2012) Shark skin effect. In: Bhushan B (ed) Encyclopedia of nanotechnology. Springer, Netherlands, pp 2400–2411. doi:10.1007/978-90-481-9751-4_159

Biomimicry Institute (2007) Biomimicking sharks. http://biomimicryinstitute.org/home-page-content/home-page-content/biomimicking-sharks.html. Accessed 27 July 2013

Bixler GD, Bhushan B (2012) Biofouling: lessons from nature. Philos Trans R Soc A 370(1967):2381–2417. doi:10.1098/rsta.2011.0502

Brooks BD, Brooks AE, Grainger DW (2013) Antimicrobial medical devices in preclinical development and clinical use. In: Moriarty TF, Zaat SAJ, Busscher HJ (eds) Biomaterials associated infection: immunological aspects and antimicrobial strategies. Springer, New York, pp 307–355. doi:10.1007/978-1-4614-1031-7_13

Chapman J, Hellio C, Sullivan T, Brown R, Russell S, Kiterringham E, Le Nor L, Regan F (2014) Bioinspired synthetic macroalgae: examples from nature for antifouling applications. Int Biodeter Biodegrad 86(Part A):6–13. doi:10.1016/j.ibiod.2013.03.036, Elsevier B.V., Amsterdam

Chung KK, Schumacher JF, Sampson EM, Burne RA, Antonelli PJ, Brennan AB (2007) Impact of engineered surface microtopography on biofilm formation of Staphylococcus aureus. Biointerphases 2(2):89–94. doi:10.1116/1.2751405

Costerton JW, Stewart PS, Greenberg EP (1999) Bacterial biofilms: a common cause of persistent infections. Science 284(5418):1318–1322. doi:10.1126/science.284.5418.1318

Dafforn KA, Lewis JA, Johnston EL (2011) Antifouling strategies: history and regulation, ecological impacts and mitigation. Mar Pollut Bull 62(3):453–465. doi:10.1016/j.marpolbul.2011.01.012

Flemming HC (2002) Biofouling in water systems—cases, causes and countermeasures. Appl Microbiol Biotechnol 59(6):629–640. doi:10.1007/s00253-002-1066-9

Florida Museum of Natural History (2013) Ichthyology at the Florida Museum of Natural History: Basking Shark Biological Profile. http://www.flmnh.ufl.edu/fish/gallery/descript/baskingshark/baskingshark.html. Accessed 26 Sept 2013

Friedmann E, Portl J, Richter T (2010) A study of shark skin and its drag reducing mechanism. In: Rannacher R, Sequiera A (eds) Advances in mathematical fluid mechanics. Springer, Berlin/Heidelberg, pp 271–285. doi:10.1007/978-3-642-04068-9_16

Gilbert PW (1984) Biology and behaviour of sharks. Endeavour 8(4):179–187. doi:10.1016/0160-9327(84)90082-6

Glinel K, Thebault P, Humblot V, Pradier CM, Jouenne T (2012) Antibacterial surfaces developed from bio-inspired approaches. Acta Biomater 8(5):1670–1684. doi:10.1016/j.actbio.2012.01.011

Harris LG, Richards RG (2004) Staphylococcus aureus adhesion to different treated titanium surfaces. J Mater Sci Mater Med 15(4):311–314. doi:10.1023/B:JMSM.0000021093.84680.bb

Lang AW, Motta P, Hidalgo P, Westcott M (2008) Bristled shark skin: a microgeometry for boundary layer control? Bioinspir Biomim 3:1–9. doi:10.1088/1748-3182/3/4/046005

Lang AW, Habegger ML, Motta P (2012) Shark skin drag reduction. In: Bhushan B (ed) Encyclopedia of nanotechnology. Springer, Netherlands, pp 2394–2400. doi:10.1007/978-90-481-9751-4_266

Lee K (2009) Rough scales make smooth sails. http://planet.wwu.edu/PDFs/2009_winter.pdf. Accessed 27 July 2013

Magin CM, Cooper SP, Brennan AB (2010) Non-toxic antifouling strategies. Mater Today 13(4):36–44. doi:10.1016/S1369-7021(10)70058-4

Maki DG, Tambyah PA (2001) Engineering out the risk for infection with urinary catheters. Emerg Infect Dis 7(2):342–347. doi:10.3201/eid0702.700342

Malshe A, Rajurkar K, Samant A, Hansen HN, Bapat S, Jiang W (2013) Bio-inspired functional surfaces for advanced applications. CIRP Ann-Manuf Technol 62(2):607–628. doi:10.1016/j.cirp.2013.05.008, Elsevier B.V., Amsterdam

Naresh MD, Arumugam V, Sanjeevi R (1997) Mechanical behavior of shark skin. J Biosci 22(4):431–437. doi:10.1007/BF02703189

National Geographic (2013) The real megalodon: prehistoric shark behind doc uproar. http://news.nationalgeographic.com/news/2013/08/130807-discovery-megalodon-shark-week-great-white-sharks-animals/. Accessed 26 Sept 2013

Oeffner J, Lauder GV (2012) The hydrodynamic function of shark skin and two biomimetic applications. J Exp Biol 215:785–795. doi:10.1242/jeb.063040

Pogoreic D (2012) Texture of shark's skin inspires a unique approach to bacteria control for healthcare. http://medcitynews.com/2012/12/texture-of-sharks-skin-inspires-a-unique-approach-to-bacteria-control-for-healthcare/. Accessed 27 July 2013

Public Policy Polling (2013) Animals and pets poll: Americans prefer dogs; fear snakes. http://www.publicpolicypolling.com/main/2013/06/animals-and-pets-poll-american-prefer-dogs-fear-snakes.html. Accessed 26 Sept 2013

Reddy ST, Chung KK, McDaniel CJ, Darouiche RO, Landman J, Brennan AB (2011) Micropatterned surfaces for reducing the risk of catheter-associated urinary tract infection: an in vitro study on the effect of Sharklet micropatterned surfaces to inhibit bacterial colonization and migration of uropathogenic Escherichia coli. J Endourol 25(9):1547–1552. doi:10.1089/end.2010.0611

ReefQuest Centre for Shark Research (2003) Biology of sharks and rays: skin of the teeth. http://www.elasmo-research.org/education/white_shark/scales.htm. Accessed 10 Aug 2013

Singh AV, Rahman A, Sudhir Kumar NVG, Aditi AS, Galluzzi M, Bovio S, Barozzi S, Montani E, Parazzoli D (2012) Bio-inspired approaches to design smart fabrics. Mater Des 36:829–839. doi:10.1016/j.matdes.2011.01.061

Southall EJ, Sims DW (2003) Shark skin: a function in feeding. Proc R Soc Lond 270:S47–S49. doi:10.1098/rsbl.2003.0006

Stewart PS, Costerton JW (2001) Antibiotic resistance of bacteria in biofilms. Lancet 358:135–138. doi:10.1016/S0140-6736(01)05321-1

Wainwright SA, Vosburgh F, Hebrank JH (1978) Shark skin: function in locomotion. Science 202(4369):747–749. doi:10.1126/science.202.4369.747

Mother-of-Pearl: An Engineering Gem

Michelle Lee

Ancient Creatures with Modern Day Applications

Imagine a world where the majority of North America lay below the equator with a southern, tropical climate. Though some people living in Chicago and other places where the weather often seems to have its own agenda may like the thought of a dependably warm climate, it would, for the most part, be very strange. Now, imagine if Scandinavia, European Russia, and eastern Europe lay to the south, east Asia and China lay in fragments, and Africa, Antarctica, India, and South America were all joined together in a massive clump of land. To top it all off, imagine this world with no plants and vegetation at all.

As unfamiliar and strange as it seems, the above description was actually the state of the Earth about 545 million years ago during what is known as the Cambrian Period (University of California Museum of Paleontology 2011). During our day-to-day lives, it rarely crosses our mind that the world we live in existed before—*way* before—time as we know it. It is easy to forget that we modern humans have only lived on this Earth for a mere blip in time—an estimated 200,000 years (Natural History Museum 2013)—compared to that of the history of the Earth. Evolution has been at work long before our arrival, crafting and creating life forms that endure the test of time.

Mollusca encompass many of these ancient life forms. They are a phylum of animals composed of at least 50,000 living species and nearly 200,000 total (Bunje 2003), first appearing on Earth about 545 million years ago at the beginning of the Cambrian Period. Species include conch shells, abalone, clams, mussels, and oysters, among many more. While the outer coverings of Mollusca are usually hard shells,

M. Lee (✉)
Mechanical Engineering, McCormick School of Engineering
at Northwestern University, Evanston, IL 60208, USA
e-mail: MichelleLee2013@u.northwestern.edu

M. Lee (ed.), *Remarkable Natural Material Surfaces and Their Engineering Potential*,
DOI 10.1007/978-3-319-03125-5_3, © Springer International Publishing Switzerland 2014

their bodies are soft. Around 440–500 million years ago, the size of Mollusca increased greatly from their original 2–5 mm sized bodies, and, at the same time, appeared in numerous classes. Despite these changes, the shells of ancient Mollusca, which will be the focus of this chapter, are generally similar to the ones of today (Espinosa et al. 2009).

Molluscs and Their Shells

Purposes of the Shell

Mollusc shells have quite a bit of pressure when it comes to what they must accomplish. As the primary barrier between a Mollusc's soft body and the outer environment, these shells must provide protection to damage from predators, remain intact without shattering under tidal waves, and resist pressures around hydrothermal vents found in the deep ocean where many Molluscs reside. At the same time, Mollusc shells must allow for the intake of nutrients, elimination of waste matter, and other natural processes such as reproduction and growth (Kaplan 1998). Shells also anchor, dig in mud, camouflage, and even grow seaweeds in its spine for the consumption of the housed mollusk (Akella 2012). Despite such a long list of responsibilities, nature has fine tuned the Mollusc shell over millions of years to develop a shell that not only meets these requirements but also displays a set of enviable mechanical properties that fascinates and inspires researchers today (Zhu and Barthelat 2011).

Composition and Types

Molluscan shells are grown by a soft tissue called the mantle that lines the inside of the shell. From this process, numerous types of shells have developed, such as foliated and cross lamellar, prismatic, and columnar and sheet nacreous structures (Espinosa et al. 2009). Regardless of type, the shells of Molluscs are generally composed of 95–99 % aragonite crystals (calcium carbonate) and 1–5 % organic matter (Kaplan 1998).

Despite common makeup, the types of shells differ in mechanical properties, with nacreous structures coming out on top with the highest strength: about 120 MPa, compared to strengths of around 60 MPa among non-nacreous structures (Espinosa et al. 2009). For this reason, nacreous shells have become very popular among researchers, and many studies have confirmed the outstanding mechanical properties of the material. This chapter will highlight these properties, as well as explore the structural reasons for such excellence.

Nacreous Shell Morphology: A Strong, Hierarchical Structure

First Level of Hierarchy

Nacreous shells, like many materials in nature—some of which have already been mentioned in this book—consist of a hierarchical structure, meaning that particular structures exist at distinct length scales. At the first level of hierarchy, which is on the millimeter scale, nacreous shells resemble an armor system composed of two levels: a hard layer of large calcite crystals on the outside and a more ductile, inner layer called nacre.

Though difficult to penetrate mechanically (for example, due to predator aggression), the outer calcite layer is prone to brittle failure. By contrast, nacre is relatively ductile and protects the soft Mollusc body while retaining structural integrity, even in the presence of cracks (Espinosa et al. 2009). This composite armor system is optimal, because the outer layer protects from penetration while mechanical energy experiences inelastic dissipation through the nacreous layer (Luz and Mano 2009).

Nacre, which is commonly known as the lustrous mother-of-pearl (Kaplan 1998) as shown in Fig. 3.1, lines the inside of many shells and is also what pearls are made of (Luz and Mano 2009). It exhibits immense strength and toughness, though it is primarily composed of aragonite (or calcium carbonate crystals), a fragile ceramic (Espinosa et al. 2009) that, on its own, is very hard and brittle (Denkena et al. 2010). The conventional definition for ceramics is a compound between non-metallic and metallic elements. In history, ceramics have come to attain their desired state via firing, a heat treatment process during which ceramics are exposed to very high temperatures. In fact, the term 'ceramic' is derived from 'keramikos', which is the Greek word for burnt stuff. Popular ceramics that are encountered in every day life include glass, porcelain, bricks, and china, and ceramics such as silicon carbide and magnesium oxide can be found in many applications such as automobiles and electronics. Nacre is also considered a ceramic, because it is a compound of non-metallic and metallic elements. However, nacre is unlike fired ceramics that have low energy absorption and are very brittle (Akella 2012). Instead, the strength of nacre is one of the highest among shells, and its fracture toughness is about 3 orders of magnitude higher than materials of a single ceramic (Wang et al. 2001). While engineered ceramics are characterized by significant reductions in tensile strength compared to their constituents, nacre is characterized by a higher tensile strength than its constituents. In other words, as expressed so well by Kiran Akella, a scientist at the Defence Research and Development Organisation lab at Pune specializing in the design of ceramic-composite armour, "nature uses a weak raw material and makes a strong and tough structure, whereas we use a very strong raw material, subject it to an expensive manufacturing process involving high temperature and pressure and create a much weaker structure. This is a great lesson in effective utilization of materials that we still have to learn and adapt" (Akella 2012).

Fig. 3.1 Nacre, otherwise known as mother-of-pearl (Photo by Roy Kaltschmidt, Berkeley Lab)

Nacre's Brick-and-Mortar Architecture

The secret to nacre's superior mechanical properties lies in its layered structure. Nacreous shells' second level of hierarchy consists of the brick-and-mortar architecture of nacre (Espinosa et al. 2009). In this arrangement, the bricks are comprised of aragonite crystals and are about 500 nm thick, while the mortar is made of highly cross-linked protein and is about 30 nm thick (Lenau and Hesselberg 2013).

Despite the fact that the organic proteinaceous mortar layer only constitutes a few percent of the total weight of the shell (Kaplan 1998), it is extremely significant to the overall mechanical properties of the shell (Kaplan 1998). In the event of a crack in the nacre, the layers of protein act as energy absorbers, ensuring that the shell can withstand repeated attacks. By deforming elastically and distributing the energy to form many other micro-cracks in neighboring bricks of aragonite, a single crack propagation that would otherwise occur in a homogenous material is prevented, avoiding entire failure of the shell (Lenau and Hesselberg 2013). Furthermore, this even

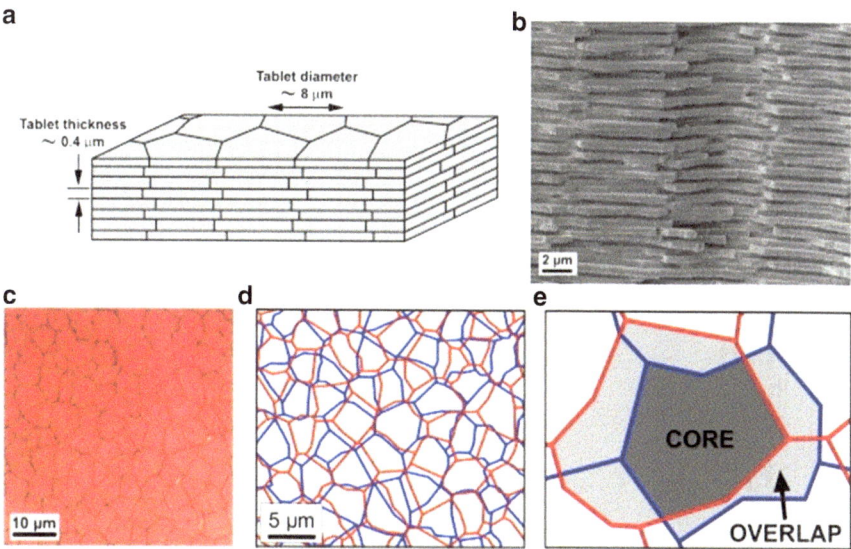

Fig. 3.2 Structure of nacre: (**a**) arrangement of tablets in nacre; (**b**) fracture surface of nacre as seen under scanning electron microscope; (**c**) view of tiling pattern of nacre from the top; (**d**) schematic of one layer to the next; (**e**) arrangement of core and overlap areas in nacre layers (Reprinted from Barthelat et al. (2007). With permission from Elsevier)

distribution of energy prevents force peaks from occurring and causing failure, increasing hardness and strength (Denkena et al. 2010).

The bricks are made up of even more layers, comprising the next level hierarchy at the micro-scale. Each layer is an aragonite polygonal tablet, with thicknesses ranging from 0.2–1.5 μm and average diameters of 5–20 μm (Denkena et al. 2010). The arrangement of tablets vary among species of Mollusca, with abalone and other gastropods having columnar nacre, where tablets are arranged in columns, while oysters, mussels, and other bivalves have sheet nacre, where tablets are arranged randomly. However, in both columnar and sheet nacre, the size and arrangement of tablets are uniform (Espinosa et al. 2009). Figure 3.2 shows an arrangement of tablets in nacre and a fractured surface, as well as the top view of the polygonal-shaped tablets with highlighted overlapping. Note: the nacre shown in Fig. 3.2 is columnar.

Sliding Between Layers of Tablets

The layered structure of tablets and soft protein also enhances the mechanical properties of nacre by allowing sliding. Nacre's high toughness can be explained in part by the material's ability to extend up to 2 % during tensile testing, which is a result of the nano-space between tablets occupied by the organic protein matrix that allows them to slide up to 100–200 nm. In this case, failure occurs when a tablet is removed due to sliding (Denkena et al. 2010).

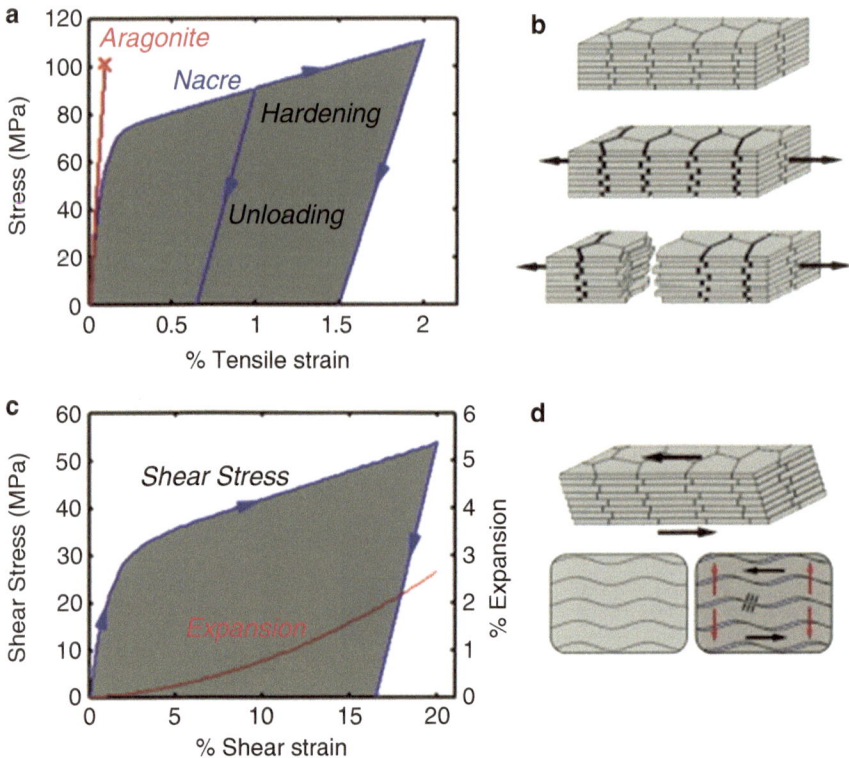

Fig. 3.3 (**a**) Experimental stress-strain curve for sample under tensile stress and (**b**) resulting deformation. (**c**) Experimental stress-strain curve for sample under shear stress and (**d**) resulting deformation (Reprinted from Espinosa et al. (2009). With permission from Elsevier)

This sliding ability enables nacre to exhibit failure strains of ductile materials, despite its composition being 95 % aragonite. The contrast in behavior between pure aragonite and the composite material of nacre under tensile strain is visible in Fig. 3.3a, where the aragonite displays abrupt, brittle failure at relatively small strains compared to the large failure strains of nacre. Under a tensile stress of about 60 MPa, the tablets slide as the interfaces between them yield in shear. This happens locally until the deformation eventually spreads to greater areas, showing large deformations and strains at the macro-scale. Finally, at sliding distances of 100–200 nm, failure takes place when the tablets pull out from each other, as shown in the progression in Fig. 3.3b. The stress-strain curve for a tablet under shear stress and the method of resistance due to waviness can be seen in Fig. 3.3c, d (Espinosa et al. 2009).

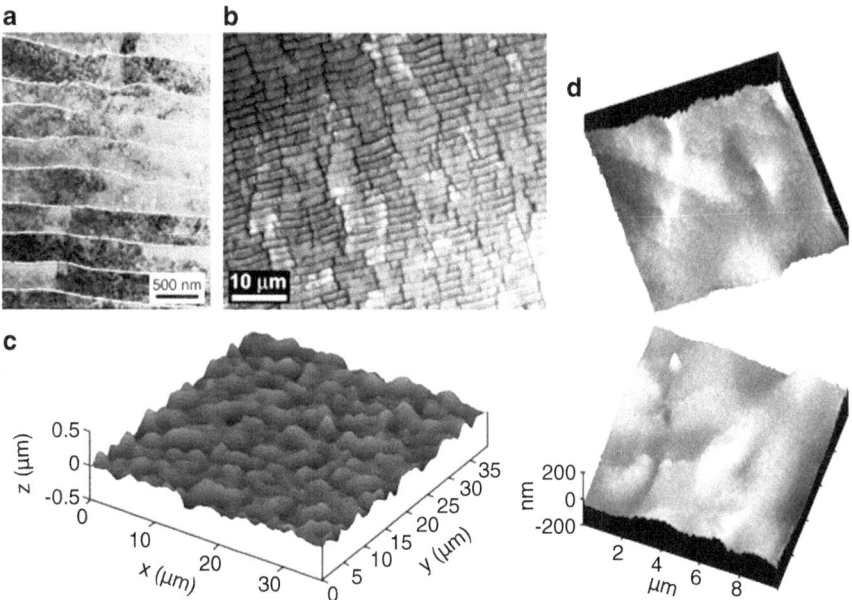

Fig. 3.4 (**a**) Transmission electron micrograph of nacre at micro-scale level reveals tablet waviness; (**b**) nacre sample from fresh water mussel *Lampsilis cardium* as seen in optical micrograph; (**c**) laser profilometry showing surface of layer; (**d**) conformality of waviness of two opposed tablets confirmed by atomic force micrographs (Reprinted from Barthelat et al. (2007). With permission from Elsevier)

Tablet Waviness

Another micro-scale feature of the tablets is a waviness that exists on many Mollusca species, such as the top shell (*Trochus niloticus*) and freshwater mussel (*Lampsilis cardium*), and that can be seen using methods such as scanning probe microscopy, optical microscopy, and scanning and transmission electron microscopy. Through these observations, it has been found that the tablets are conformal in their waviness, fitting together perfectly (Espinosa et al. 2009). Upon execution of roughness analysis on nacre from fresh water mussel *Lampsilis cardium*, shown in Fig. 3.4, Barthelat et al. found that the surface had a root mean square (RMS) of 85 nm for an average distance of 3 μm peak-to-peak. The amplitude of the roughness was significant, exceeding 200 nm for tablets with an average thickness of 450 nm, showing that the waviness was very prominent (Barthelat et al. 2007).

To analyze and quantify the effect of waviness on nacre's behavior upon crack formation, Espinosa et al. created models of a flat tablet and wavy tablet based on actual tablet contour and then increased the applied J. This simulation revealed a highly apparent difference between the crack propagation of the flat and wavy models, which is evident in Fig. 3.5. While the flat model showed a crack that was generally localized in the crack plane, the wavy model distributed the inelastic deformation

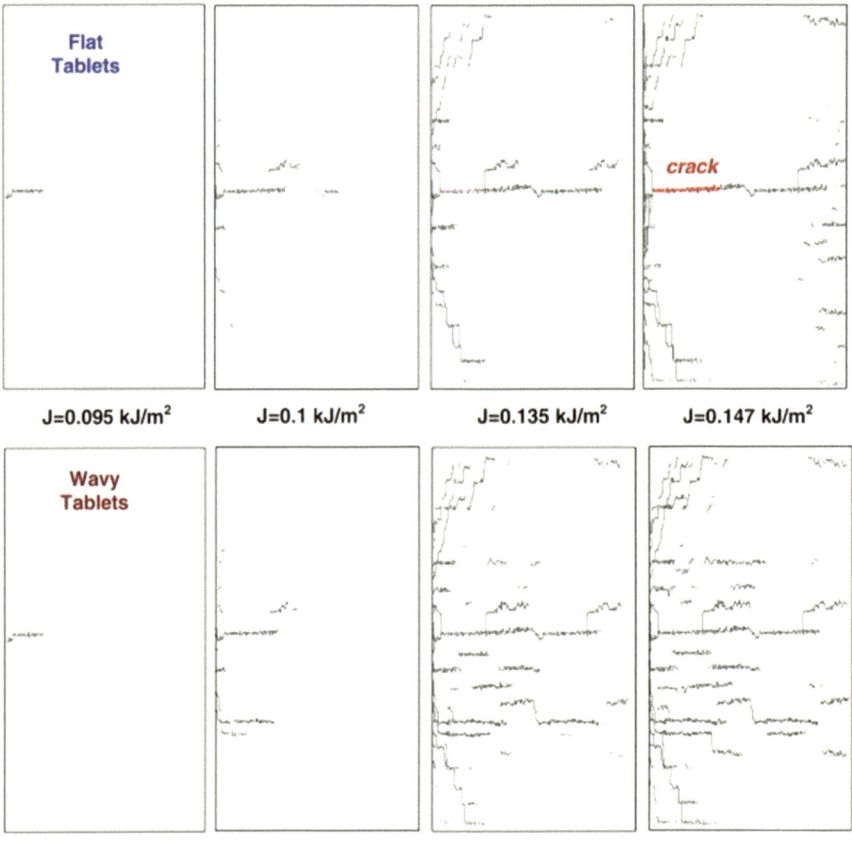

Fig. 3.5 Comparison of flat and wavy tablet models in their response to an increasing applied *J. Black lines* indicate displacement jumps over 20 nm, with jumps over 600 nm indicated by 'crack' (Reprinted from Espinosa et al. (2009). With permission from Elsevier)

over a much larger region. As a result, the flat tablet ultimately cracked (or reached a displacement jump of over 600 nm) as shown in the last frame of Fig. 3.5, while the wavy tablet did not, emphasizing the importance of waviness in distributing inelastic deformations so as to prevent failure (Espinosa et al. 2009).

Interlocking Mechanisms Between Tablets

In addition to waviness, tablets were observed to have interlocking mechanisms shown in Fig. 3.6, rather than simply being set down on each other. The interlocking of tablets encourages deformation and progressive failure, increasing toughness and reducing the chance of catastrophic failure. Katti and Katti observed the

Fig. 3.6 Fractured surfaces of nacre examined under scanning electron microscope reveals: (**a**) organic phase characterized by large deformations, (**b**) tablet pullout, and (**c**) interlocking mechanisms between tablets (Reprinted from Katti and Katti (2006). With permission from Elsevier)

behavior of tablets' interlocking mechanism by creating finite element models. Upon simulation at an applied stress of 14 MPa, the model without interlocks showed plastic flow in the organic protein layer, indicating failure. On the other hand, when the interlocked model experienced an applied load of 26 MPa, only the interlocks nearest the load location failed. Then, as the load progressively increased, other interlocks in the area failed, until the nacre failed at a stress of 50 MPa and strain of 0.005 when all interlocks failed (Katti and Katti 2006). The results of the simulation can be seen in Fig. 3.7.

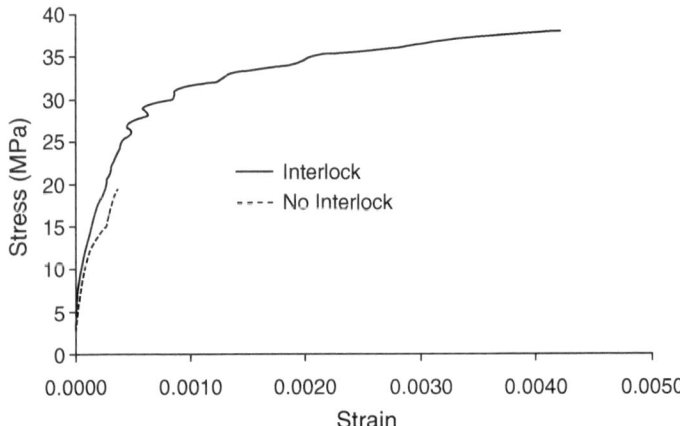

Fig. 3.7 Stress-strain responses of interlocked and non-interlocked finite element models, showing high strength and toughness in interlocked model compared to relatively early failure of non-interlocked model (Reprinted from Katti and Katti (2006). With permission from Elsevier)

Nano-asperities on Tablet Surfaces

The final hierarchical level of features discovered in nacre consists of the nano-asperities found on the surface of the aragonite tablets. These tiny aragonite grains were observed to be ductile in nature, which allows them to deform and rotate. Due to the ability for such deformation at the nano-scale level, which is encouraged by the protein biopolymer between them, the aragonite tablets are further able to dissipate energy, leading to an overall increase in fracture toughness. In addition, these asperities on the surface help to resist shear forces (Luz and Mano 2009).

Considering a major component of the makeup of nacre is hard, brittle aragonite, nacre's mechanical properties are astounding. However, nacre's super mechanical properties are not surprising, particularly after taking a look at its many structural strategies for toughening and strength. Scanning electron microscope images of fractured nacre samples show the features that make the mechanical response to stress inherent to the material, and these include deformations in the organic matrix during sliding, the pullout of tablets, and the interlocking mechanisms between tablets (Katti and Katti 2006), which have all been discussed in this section. Other factors such as waviness and nano-asperities enhance such superior mechanical responses, making nacre truly a one-of-a-kind composite.

Engineering Applications

Nacre-Inspired Nano-composites

The superior mechanical properties of nacre have been confirmed by numerous studies, establishing its identity as a strong inspiration for novel layered nano-composites (Luz and Mano 2009). According to Tang et al., "finding a synthetic pathway to artificial analogs of nacre and bones represents a fundamental milestone in the development of composite materials." In 2003, Tang et al. succeeded in creating a nano-scale version of nacre using organic and inorganic layers consisting of polyelectrolytes and clays. Their material exhibited a tensile strength close to that of nacre and a Young's modulus close to that of lamellar bone (Tang et al. 2003).

Tang et al. were not the only ones to create synthetic composites resembling nacre. In 2011, Zhu and Barthelat created a prototype of a nacre-like material composed of poly-methyl-methacrylate (PMMA) tablets. They assembled this synthetic material by layering PMMA tablets in a columnar arrangement, simply allowing direct contact and dry friction at the interfaces to dictate the interaction between layers. To mimic the mechanisms of nacre, Zhu and Barthelat introduced waviness to the tablets, as well as transverse fasteners to reinforce the composite.

Subsequent analyses of the material under uniaxial tension in the direction of the tablets confirmed their success in duplicating nacre's mechanisms, including tablet interlock and the dissipation of energy throughout the material through distribution of deformation. Though the material had a large strain capacity—about 9–12 %—it was unable to withstand large tension forces, making it unsuitable for use in current engineering applications. However, the analytical and finite element models that were created based on this material can be used in the future to attain optimal combinations of toughness, strength, and modulus that, they hope, can be applied to real-world engineering endeavors (Zhu and Barthelat 2011).

Bone Implants

Though researchers like Tang, Zhu, and Barthelat strive to produce synthetic composites that display excellent mechanical properties because of their nacre-like structures, others are looking at the possibility of using nacre itself in a location of utmost importance: human bodies. The implants that are available today for use in the human body are lacking in several ways. Their life times are unsatisfactory, and they frequently wear, sometimes causing inflammatory reactions due to wear particles. These inflammatory reactions can even cause the prosthesis-bone interface to experience premature loosening. Nacre's biocompatibility has made it available to scientists as a possible alternative to existing implant materials (Denkena et al. 2010).

To test nacre as a bone implant, Atlan et al. inserted solid nacre into the femurs of 16 sheep. Remarkably, all of the sheep healed with no clinical problems or infections in the nacre-bearing limb. After 3 months, the bone and tissue of the femur grew into the nacre implant, so that the interface between them was bonded in several locations. At this point, the nacre was anchored to the bone as a part of it. After 10 months, there was still no indication of inflammation or general intolerance. Furthermore, discovery of a cellular bed around the nacre implant that leads to an osteoprogenitor layer indicated that the nacre implant triggers and encourages osteogenic—or bone forming—activity. Given the biocompatibility of nacre and bone as well as the durability and mechanical properties of nacre, it can definitely be considered as a potential replacement for defective bone (Atlan et al. 1999).

Based on the sheer amount of useful applications and innovations that nacre has bioinspired, mother-of-pearl truly stands out as an engineering gem. The ingenuity of nature is made evident in the structure of nacre, where hierarchy and other structural mechanisms are utilized to form a strong, tough shell unrivaled by manmade materials. Nature has crafted the ancient mollusca over years and years, providing us with a superior natural composite that can serve as an inspiration in the design of our own novel composites.

References

Akella K (2012) Biomimetic designs inspired by seashells. Resonance 17(6):573–591. doi:10.1007/s12045-012-0063-2

Atlan G, Delattre O, Berland S, LeFaou A, Nabias G, Cot D, Lopez E (1999) Interface between bone and nacre implants in sheep. Biomaterials 20(11):1017–1022. doi:10.1016/S0142-9612(98)90212-5

Barthelat F, Tang H, Zavattieri PD, Li CM, Espinosa HD (2007) On the mechanics of mother-of-pearl: a key feature in the material hierarchical structure. J Mech Phys Sol 55(2):306–337. doi:10.1016/j.jmps.2006.07.007

Bunje P (2003) The Mollusca. http://www.ucmp.berkeley.edu/taxa/inverts/mollusca/mollusca.php. Accessed 31 July 13

Denkena B, Koehler J, Moral A (2010) Ductile and brittle material removal mechanisms in natural nacre—a model for novel implant materials. J Mater Proc Technol 210(14):1827–1837. doi:10.1016/j.jmatprotec.2010.06.014

Espinosa HD, Rim JE, Barthelat F, Buehler MJ (2009) Merger of structure and material in nacre and bone—perspectives on de novo biomimetic materials. Prog Mater Sci 54(8):1059–1100. doi:10.1016/j.pmatsci.2009.05.001

Kaplan DL (1998) Mollusc shell structures: novel design strategies for synthetic materials. Curr Opin Solid State Mater Sci 3(3):232–236. doi:10.1016/S1359-0286(98)80096-X

Katti KS, Katti DR (2006) Why is nacre so tough and strong? Mat Sci Eng C 26(8):1317–1324. doi:10.1016/j.msec.2005.08.013

Lenau T, Hesselberg T (2013) Biomimetic self-organization and self-healing. In: Lakhtakia A, Martín-Palma RJ (eds) Engineered biomimicry. Elsevier B.V, Amsterdam. doi:10.1016/B978-0-12-415995-2.00013-1

Luz GM, Mano JF (2009) Biomimetic design of materials and biomaterials inspired by the structure of nacre. Philos Trans Math Phys Eng Sci 367(1893):1587–1605. doi:10.1098/rsta.2009.0007

Natural History Museum (2013) How long have we been here? http://www.nhm.ac.uk/nature-online/life/human-origins/modern-human-evolution/when/index.html. Accessed 26 Aug 2013

Tang Z, Kotov NA, Magonov S, Ozturk B (2003) Nanostructured artificial nacre. Nat Mater 2:413–418. doi:10.1038/nmat906

University of California Museum of Paleontology (2011) The Cambrian Period. http://www.ucmp.berkeley.edu/cambrian/cambrian.php. Accessed 26 Aug 2013

Wang RZ, Suo Z, Evans AG, Yao N, Aksay IA (2001) Deformation mechanisms in nacre. J Mater Res 16(9):2485–2493. doi:10.1557/JMR.2001.03400

Zhu D, Barthelat F (2011) A novel biomimetic material duplicating the structure and mechanics of natural nacre. In: Proulx T (ed) Mechanics of biological systems and materials, vol 2. Springer, New York. doi:10.1007/978-1-4614-0219-0_25

Diatoms: Glass Ornaments of the Earth's Waters

Mindie Chu

Miniature One-of-a-Kind Wonders

The first snowfall of the season is always the best. By this time of year, everyone is tired of the dreary, rainy weather, and although still cold, the beautiful, pristine snow brings back the joy of the season and serves as a reminder of the holiday season that is to come. When the first snowfall comes, children in the neighborhood rush out of the house and run around in their puffy jackets, gloves, and rain boots with their tongues sticking out as they try to catch snowflakes in their mouths. Parents tell their kids, "You know, if you look really closely, every snowflake is different!" Indeed, the event of a repeat in snowflake geometry is mathematically impossible based on the endless combinations of micro-weather conditions during each snowflake formation (Armitage 1995). Just as snowflakes are all unique, like the ones in Fig. 4.1, the Earth's waters contain hundreds of thousands of different species of beautiful, microscopic glass crystals called diatoms.

Among every single one of its many species, diatoms exhibit exquisite architecture of their shells. The astonishing beauty of diatom shells has captivated artists, scientists, and researchers, especially due to the non-adaptational quality of their geometric diversity. In other words, the beautiful shapes and patterns on their shells have not been determined by natural selection, making diatoms an example of "beauty for its own sake, unfettered by evolutionary pragmatism," just like snowflakes. In addition to being a visual delight, diatoms play a critical part in the ecology of the Earth (Armitage 1995), and their applications in engineering are growing ever more known in the scientific community.

M. Chu (✉)
Mechanical Engineering, McCormick School of Engineering
at Northwestern University, Evanston, IL 60208, USA
e-mail: MindieChu2013@u.northwestern.edu

M. Lee (ed.), *Remarkable Natural Material Surfaces and Their Engineering Potential*,
DOI 10.1007/978-3-319-03125-5_4, © Springer International Publishing Switzerland 2014

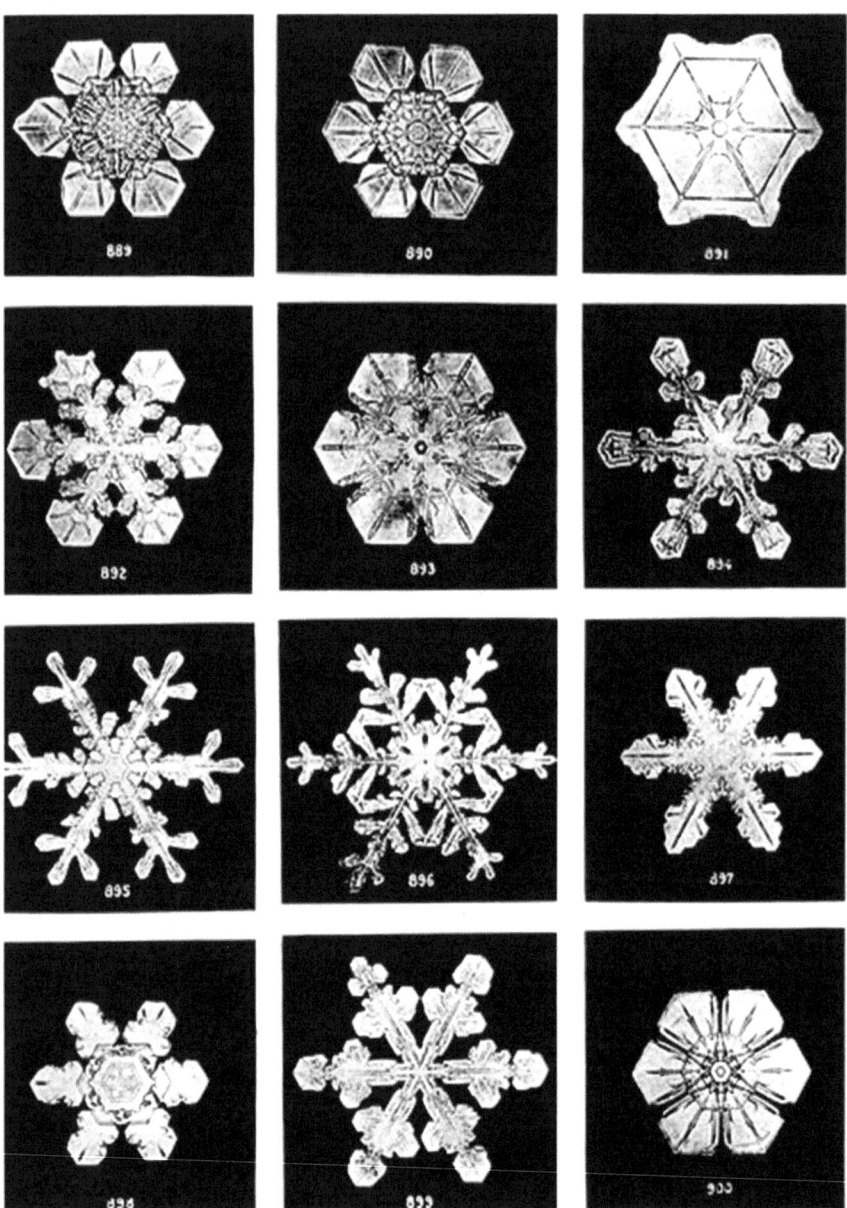

Fig. 4.1 Photographs of individual snowflakes taken in 1902 by Wilson Bentley, "The Snowflake Man." Plate XIX of "Studies among the Snow Crystals…" from the Annual Summary of the "Monthly Weather Review" (Publication of the U.S. Department of Commerce, National Oceanic & Atmospheric Administration (NOAA) (2013a) Photo Library)

What Are Diatoms?

Composition and Environment

Diatoms have been studied for many years, and early illustrations of diatoms by Ernst Haeckel, a German biologist, philosopher, and physicist, were published in his book *Kunstformen der Natur (Art Forms in Nature)* in 1904. One of his many illustrations, which can be seen in Fig. 4.2, depicts the variety and sophistication of the shapes and forms of diatom frustules (or shells) that he encountered in studying diatoms (Marine Biological Laboratory Woods Hole Oceanographic Institution 2013).

Diatoms, as shown in Fig. 4.3, are so numerous that they make up about 90 % of the ocean's living organisms (Armitage 1995). Diatoms are one of the most common microaquatic single celled algae. They are eukaryotic and photosynthetic and have an estimated 200,000 species and 250 living genera. As microscopic organisms, their sizes range from 2 μm to 2 mm (Gordon et al. 2008). So plentiful and ubiquitous in the earth's environment, scientists believe them to account for around 25 % of the net amount of organic carbon produced in the entire world (Parkinson and Gordon 1999). They are housed in silica frustules or shells and utilize yellow-brown chloroplasts to synthesize glucose through photosynthesis from nutrients and resources in their environment. In addition to the green chlorophyll pigments that exist in the leaves of many plants we see every day, diatom chloroplasts also contain beta-carotene (orange), fucoxanthin (olive green/brown), diatoxanthin (yellow), and diadinoxanthin (yellow), which, when combined, give the diatoms their yellow-brown color. Depending on the concentration of each pigment, diatoms vary in color from green to dark brown (Van Egmond 1995).

Diatoms exist in any body of water that has enough nutrients—oceans, lakes, rivers, ponds, and even in household aquariums. They exist in various forms: planktonic or free-floating, colonial or solitary, or attached to objects such as sea ice, rocks, or other algae (Leventer 2009). Because they rely on the sun for energy, they can only exist in the photic zone, which is as deep as the sunlight can penetrate the water and differs depending on the clarity of water. Some are free-living species that rely on water currents and other environmental variables to stay afloat in the photic zone of water. Others are benthic and can survive at the bottom of shallow pools of water. In some areas, benthic diatoms do not produce as much photosynthetic product as free-living or free-floating species due to the conditions of the environment (Spaulding et al. 2010). For example, the sediments of the Netherlands' Wadden Sea contain many benthic diatom flora. During high tide, however, they are not within the photic zone and thus are not able to undergo photosynthesis. Rough weather can also cause the sediment that the diatoms are fixated on to be washed away, which can remove them from the photic zone (Van den Hoek et al. 1996).

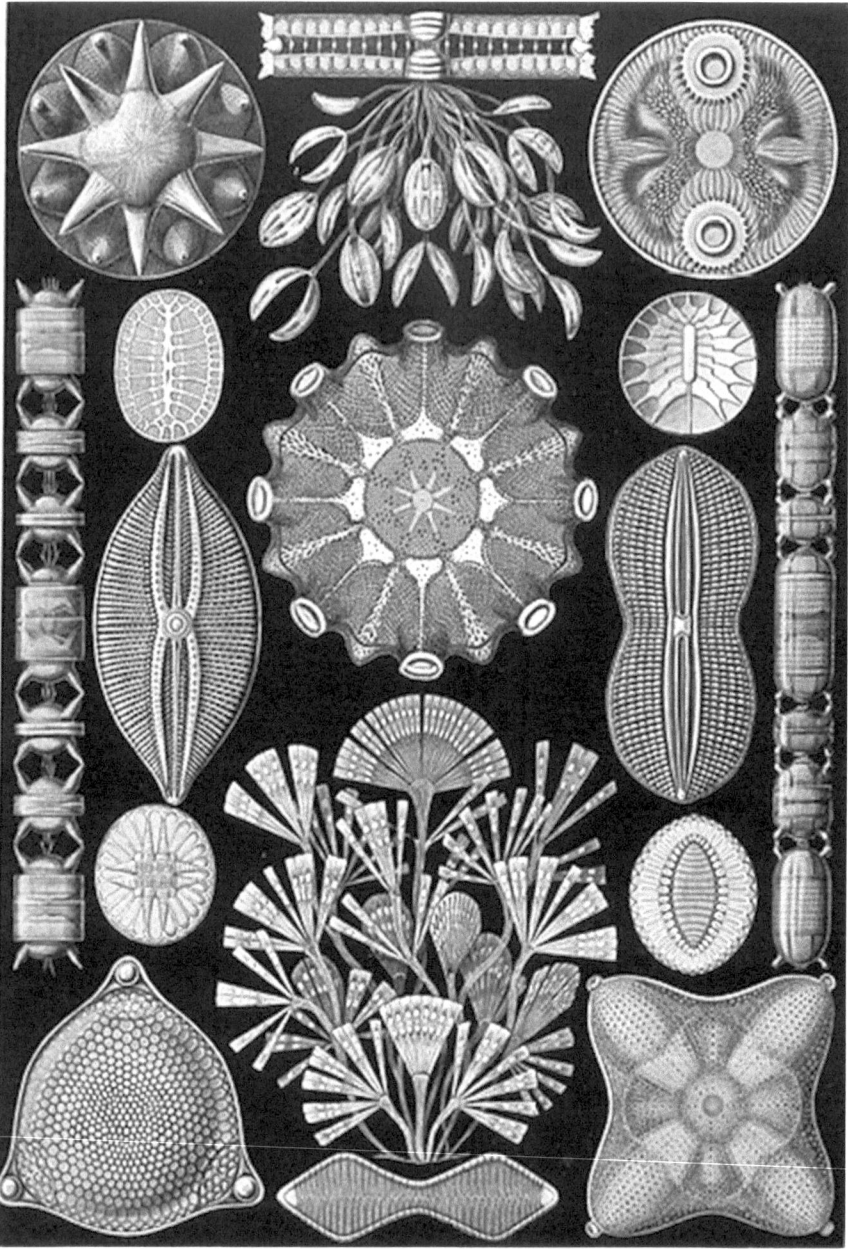

Fig. 4.2 Ernst Haeckel's early illustrations of diatoms in *Kunstformen der Natur* (Released into the public domain due to age)

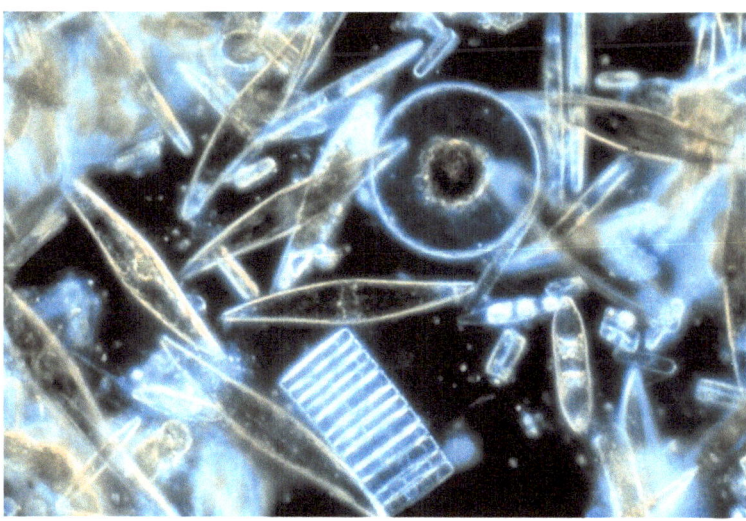

Fig. 4.3 Diatoms among crystals of sea ice in McMurdo Sound, Antarctica, taken by Professor Gordon T. Taylor of Stony Brook University. Originally from 1983 on a 35 mm Ektachrome slide. Digitized as part of National Oceanic & Atmospheric Administration (NOAA) At the Ends of the Earth Collection (U.S. Department of Commerce, National Oceanic & Atmospheric Administration (NOAA) (2013b) Photo Library)

Diatomic Structure

To the naked eye, diatoms take on the appearance of scum at the top of the ocean, lake, or even the back of a whale. The slimy brown patches present on rocks in rivers are actually layers of diatoms. While they look like muck on the rock, the cells are actually surrounded by a beautiful glass silica shell. Called frustules, the silica shells of diatoms are made from the silica present in the surrounding environment. The frustules consist of two halves: the hypotheca and epitheca, as shown in Fig. 4.4. The hypotheca can be broken down into the hypovalve and hypocingulum, and the epitheca into the epivalve and epicingulum. The cingula are also known as girdle bands of the diatom, and depending on the species, there may be more than one girdle band in the cingulum. Each hypovalve and epivalve consists of the valve face and the mantle, and the cingulum surrounds the mantle. The hypotheca and epitheca fit together like a box and lid, with the hypotheca being the box and epitheca the lid (University College London 1999). Another way to describe the frustule is using a petri dish assembly, where the epitheca is like the lid of the petri dish while the hypotheca is the petri dish itself. The tops and bottoms are the valve faces and the sides of the lid and the dish are the mantle and cingula.

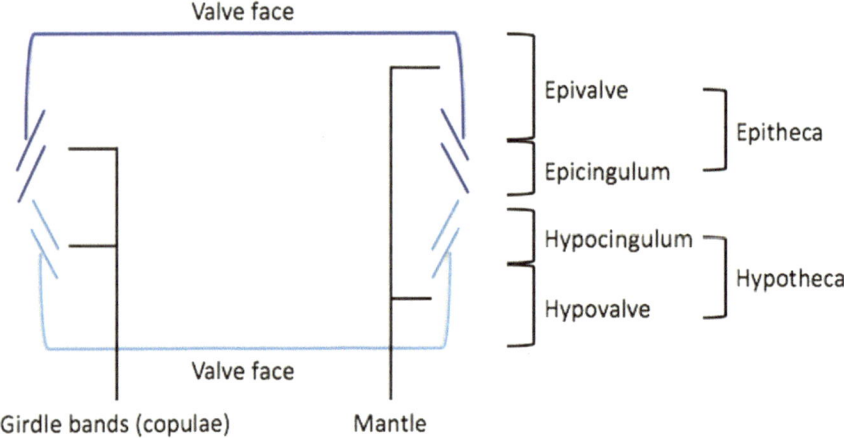

Fig. 4.4 Box-and-lid type structure of diatoms

Hierarchical Structure of Diatoms

The frustules of all diatoms are hierarchical, though there may be variations in the hierarchical structure depending on the particular species. In fact, diatomic frustules have been found to consist of several scales of characteristic lengths. For example, the diatomic taxon Conscinodiscus sp. has three major layers in its frustules: the foramen (most inward), the cribrum, and the cribellum (most outward). There is an additional honeycomb arrangement called the areola between the foramen and cribrum, and the cribrum and cribellum are characterized by repetitive dome structures on the surface. Lastly, the three major layers are porous, with the pore size of each layer distinctly different from another. For a sense of scale, the pores of the cribellum may reach a size of about 40 nm in diameter, while those of the foramen may be about 1 µm in diameter (Dimas and Buehler 2011).

Characteristics and Scientific Classification

The distinct characteristics of the silica shells of different diatom species are used to classify them into different orders, suborders, families, and species. Diatoms are in division Chrysophyta, class Bacillariophyceae. They are separated into two general orders, Centrales (also known as Biddulphiales) and Pennales (also known as Bacillariales), by the general shape of their frustules.

The petri dish analogy works exceptionally well to explain the diatoms of order Centrales since they have an overall radial symmetry. They are further separated into three suborders in relation to the location central areola on the valve faces: Coscinodiscineae for diatoms with nonpolar central areola, Rhizosoleniineae for those with unipolar central areola, and Biddulphiineae for those with bipolar symmetry.

Coscinodiscineae, which have nonpolar central areola, have a wide girdle and are thus referred to as drum-shaped, while Rhizosoleniineae have a thinner girdle and are referred to as coin-shaped. Biddulphiineae are much thicker than the other two suborders of the Centrales diatoms.

Pennales diatoms are characterized by their bilateral symmetry. They are also generally broken down into two subgroups: one with a raphe system and one without. The raphe system is a longitudinal slit along the valve that allows the diatom to secrete mucilage from a wall organelle called rimoportula (labiate process) that is used by some diatoms as a means of mobility and others as a means of fixation to a surface. The mucilage not only helps the diatoms to adhere to surfaces, but it can also form a skin to protect the diatom from erosion by waves and rough tides. This mucilage secretion also allows diatoms to adhere to each other, forming a chain of diatoms in a colony. Since no Centrales diatoms have been observed to have raphe systems, they are largely nonmotile. Similarly, Araphids do not have a raphe system and are thus nonmotile, although they may have rimoportula in their cell walls. Some raphid systems have one raphe while others have two on their valve surfaces. Some raphe systems, like those of the Nitzschioid diatoms, have raphe slits that are positioned near the edge of the valve face along a ridge (keel) and spanned internally by struts (fibulae), which appear as lines or dots in diatom images. In addition to the pores, dots, and slits, some diatoms have developed spines along the junction of the valve face and valve mantle. Such varieties in surface structures allow for easy characterization of diatoms (Van den Hoek et al. 1996).

Life Cycle

Diatoms reproduce both asexually and sexually. Asexual reproduction occurs through a process that is similar to binary fission and is common to many single-celled bacteria. During this process, the cell replicates its DNA, separates the two sets of DNA at opposite sides of the cell, and grows a new cell wall in the middle to separate the bacterium. The difference between diatom reproduction and binary fission is the separation of the two frustule halves and the glass silica shell restricting growth in cell size.

Like in binary fission, the diatom replicates its DNA and segregates each set to opposite sides of the cell. Instead of the ordinary cell wall that develops in bacteria, a new hypotheca develops for each half of the frustule, and the original frustule becomes the new epitheca of the daughter diatoms. This results in one daughter cell that is the same size as the parent, as shown on the right of the P1 row in Fig. 4.5, and one daughter cell that is slightly smaller, due to the fact that the original hypotheca, which is smaller in diameter than the epitheca, was used as the new epitheca (shown on the left in row P1 of Fig. 4.5) (University College London 1999).

As the cells continue to undergo asexual reproduction, the average cell size continues to decrease until it reaches a minimum size. At this point, the cell undergoes sexual reproduction and enters meiosis. The gametes fuse to produce an auxospore. The auxospore sheds its existing frustules and, using the nutrients and silica from the

Fig. 4.5 Diagram showing diatomic replication of DNA. Note the difference in size between two daughters, as one daughter cell uses the original hypotheca, which is smaller in diameter than the epitheca, as the new epitheca

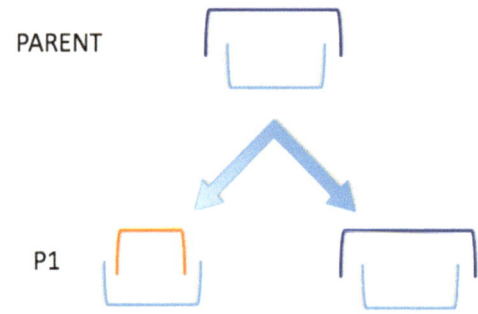

environment, grows a new frustule at the maximum size. If the necessary nutrients are not available, the cell goes into a restive state. Then, when nutrients become readily available, the cell will renew its asexual reproduction cycle in the vegetative state.

Engineering Applications

Current Uses of Diatoms

Diatoms are currently widely used in the commercial arena as well as in research. The mildly abrasive characteristics of diatom silica frustules are desirable in toothpaste and exfoliating scrubs and have thus been included in the form of diatomaceous earth, which consist of the fossils and silica skeletons of diatoms. Diatomaceous earth is also widely used as a filtration aid, absorbent for liquids, matting agent for coatings, and more. Alfred Nobel also discovered that the addition of diatomaceous earth stabilized nitroglycerin, leading to the development of dynamite.

Additionally, diatomaceous earth is useful in agriculture as an anti-caking agent for grain storage and as an insecticide. It is also used as a growing medium in hydroponic gardens since it retains water and nutrients while draining fast and freely. High oxygen circulation in the growth medium is achieved as well. Diatomaceous earth is also very useful in classifying strata in the oceans, and classifications can be made through the type and quantity of species in a certain body of water.

This material has research applications too, as it can be used for DNA purification. Diatomite and other silicates will only remove double stranded DNA but not RNA or proteins and, as such, can be used to isolate DNA from samples. The DNA can then be extracted from the diatomites with low ionic strength buffers like water. Diatoms can also be used to tell the levels of pollution in a body of water. In fact, samples prepared according to the directions written in the last section of this chapter can be used to survey the population of diatoms in a body of water. The greater the variety of species, the less polluted the water is.

Future Perspectives

Currently, the best solar panels can only absorb 30 % of the sunlight that reaches Earth. The rest is reflected back or releases as heat. In today's day and age where the search for alternate energy sources has become a portion of investments, more efficient solar cells are needed. Diatoms may be the solution to this problem. The silica-based shells of diatoms may be developed into the most efficient solar cells due to its symmetrical, complex patterns and small pores that allow light to flow into the shell without allowing any to escape. Such characteristics are exactly what researchers are looking for in solar cells.

Researchers are currently using gold, a flexible material, to make molds of different algal shells to test for the best shell structures that have optical properties for optimal absorption of solar energy. The gold molds would be used to generate a computer model that could be used for computer simulations in order to learn how light is broken up by structure, at what angle the light enters it, and how light is reflected and used within the shell. Researchers also plan on constructing the perfect model through simulation and looking for the model in nature (Norwegian University of Science and Technology 2012).

Researchers are also looking at using diatoms in the development of a nano-scale biosensor. Through gene therapy, researchers are looking to incorporate the genes that cause the diatom to express a protein containing two fluorescent proteins—one that glows blue and one that glows yellow—that will bind to the sugar, ribose. When the diatom is not in the presence of ribose, the protein will not bind to anything, and the glowing proteins will be close in proximity to one another. The energy from the blue fluorescence, which is higher due to its shorter wavelength, will be transferred to the yellow protein, causing it to glow. This is similar to the way one ball, when it hits another, will cause the other ball to bounce away as well. When the protein binds to ribose, the two fluorescing proteins are further apart, and the blue fluorescing protein will transfer less energy to the yellow fluorescing protein, causing it to glow less. In this case, more blue light will be displayed. Using the same ball analogy, if the first ball rolls into the second ball slowly and gently taps it, the second ball will not move as far. When a sensor that is covered in these glowing diatoms is exposed to ribose, the amount of each light displayed can be translated into the amount of ribose present. Researchers are trying to develop this sensor to detect threats such as explosives in the marine environment (Marshall et al. 2012).

Finally, many aspects of a diatom's shell make it a great candidate for drug delivery vehicles. For example, diatom shells have a pill-box like structure and a large, porous surface area. Other advantageous characteristics include its biodegradability and biocompatibility, as well as its easy and inexpensive cultivation. Because of these aspects, diatoms can be functionalized easily with imaging dyes, magnetic nanoparticles, biomolecule sensors, and immunotargeting bioreceptors. Even more, scientists and researchers imagine that diatom shells could be used 1 day for the bodies of self-propelled, microdevices meant to achieve complicated functions in the human body like cell repair and monitoring (Gordon et al. 2008).

Viewing Your Own Diatoms

Materials

The process of collecting and preparing your own samples is simple and straight-forward. The equipment to prepare the slides should be readily available at any school or scientific institution.

The materials necessary are as follows:

- Rock from aquarium/river/lake/etc. with brown film on top
- Plastic bin or tray
- Toothbrush
- Distilled water
- Pipette
- Test tube
- Drain cleaner
- Centrifuge tubes
- Centrifuge (manual or electronic)
- Glass slide
- Cover slip
- Hot plate
- Slide mounting media
- Tweezers

Procedure

Diatoms can be collected by any water source, like a shallow river or pond. First, look for a large rock with a flat surface that has a brown film on top. The brown film will most likely contain many diatoms. To remove the diatoms from the rock surface, place the rock in a tray or bin and scrub at the surface that was facing the sun with a toothbrush that you no longer need. After a minute, rinse the top of the rock with some water. You should see a brownish liquid run off from the surface. Continue scrubbing until you can see the rock underneath the brown film. Rinse the rock surface again. The brownish liquid that has collected at the bottom of the tray contains the diatoms you want to collect. Pipette this liquid into a vial for storage.

Because the sample is contaminated with debris from the lake and rock surface such as sediment and sand, you want to purify it with the drain cleaner. Using a pipette, take a sample from the bottom of vial, where the sediment has collected, and place in a centrifuge tube. Allow the diatoms and sediment to settle at the bottom and gently remove the liquid that is at the top, which is called the supernatant, using a pipette. Using a pipette, carefully add two to three times more drain cleaner than the diatom sample and gently pipette up and down to mix. Let the solution sit for 10 min, then gently mix again.

After letting the solution sit for another 10 min, mix again. Then dilute the solution by adding distilled water. Add distilled water to another centrifuge tube so that the two tubes are at the same water level. This is important because the centrifuge will only spin correctly if the tubes are balanced. If the centrifuge is unbalanced, the rotor may become unbalanced, and the machine may be damaged and injure the user. Centrifuge the tubes for 3 min at 1,200 rpm. Afterwards, pour out the liquid in the sample test tube into a waste bin. Repeat the dilution and centrifugation at least three more times to dilute the drain cleaner. After the dilution, the diatom cells should look like a white clump at the bottom of the centrifuge tube. Dilute the clean diatom sample with about 10 mL of distilled water.

To prepare the slide, place the cover slide on the hot plate. Pipette a few drops of the sample onto the cover slide and heat until the liquid evaporates. Using tweezers, place the cover slip so that the side with the evaporated sample is on the glass slide. Since water has a similar refraction index as glass, it may be difficult to see the diatoms. If desired, after evaporating the sample on the cover slip, place one drop of mounting media onto the glass slide and place the sample side down onto the slide. Then heat the slide on the hot plate until the liquid from the mounting media has evaporated.

Now the slide can be viewed under a microscope to observe your collected diatoms and the beauty and uniqueness of their frustules. Their applicability in the engineering realm makes diatoms and their frustules even more beautiful, pointing once again to nature, the ultimate designer and inventor.

References

Armitage MH (1995) Microgeometric design of diatoms: jewels of the sea. Acts & Facts 24(8)

Dimas L, Buehler MJ (2011) Hierarchical mechanics of diatom algae: from atoms to organism and weakness to strength. http://imechanica.org/node/11366. Accessed 27 Sept 2013

Gordon R, Losic D, Tiffany MA, Nagy SS, Sterrenburg FAS (2008) The glass menagerie: diatoms for novel applications in nanotechnology. Trends Biotechnol 27(2):116–127. doi:10.1016/j.tibtech.2008.11.003

Leventer A (2009) Diatoms. In: Gornitz V (ed) Encyclopedia of paleoclimatology and ancient environments. Springer, Netherlands, pp 279–280. doi:10.1007/978-1-4020-4411-3

Marine Biological Laboratory Woods Hole Oceanographic Institution (2013) Ernst Haeckel: art forms in nature. http://legacy.mblwhoilibrary.org/haeckel/. Accessed 21 Aug 2013

Marshall KE, Robinson EW, Hengel SM, Pasa-Tolic L, Roesijadi G (2012) FRET imaging of diatoms expressing a biosilica-localized ribose sensor. PLoS One 7(3):e33771. doi:10.1371/journal.pone.0033771

National Oceanic & Atmospheric Administration (NOAA) (2013a) NOAA Photo Library. http://www.photolib.noaa.gov/htmls/wea02087.htm. Accessed 17 Aug 2013

National Oceanic & Atmospheric Administration (NOAA) (2013b) NOAA Photo Library. http://www.photolib.noaa.gov/htmls/corp2365.htm. Accessed 17 Aug 2013

Norwegian University of Science and Technology (2012) A bright future—with algae: diatoms as templates for tomorrow's solar cells. http://www.sciencedaily.com/releases/2012/07/120717100117.htm. Accessed 10 Dec 2012

Parkinson J, Gordon R (1999) Beyond micromachining: the potential of diatoms. Trends Biotechnol 17(5):190–196. doi:10.1016/S0167-7799(99)01321-9

Spaulding SA, Lubinski DJ, Potapova M (2010) Diatoms of the United States. http://www.
westerndiatoms.colorado.edu. Accessed 10 Dec 2012

University College London (1999) Diatoms. http://www.ucl.ac.uk/GeolSci/micropal/diatom.html.
Accessed 9 Dec 2012

Van den Hoek C, Mann D, Jahns HM (1996) Algae: an introduction to phycology. Cambridge
University Press, Cambridge, UK

Van Egmond W (1995) Diatoms. http://www.microscopy-uk.org.uk/mag/indexmag.html?http://
www.microscopy-uk.org.uk/mag/wimsmall/diadr.html. Accessed 10 Dec 2012

Michelle Lee

The Lotus Flower: A Botanical Celebrity

The lotus flower (Fig. 5.1) is famous around the world across many cultures and religions. For example, it is considered to be a sacred flower by Buddhists, symbolizing cosmic harmony and spiritual illumination, as well as being symbolically equal to Buddha. It is even believed to have birthed the sun, an Egyptian myth inspired by the fact that the flower opens in the morning and closes by evening. Even more, this flower is written about extensively in Vedic and Puranic literature and is India's national flower. Many delight in the pale pink and white colors of this beautiful flower, and it is no wonder that it is known to symbolize majesty, grace, and purity, among other virtues (The Flower Expert 2005).

What many people do not know, however, is that its counterpart—the wide, flat leaf of the lotus plant—albeit not as bright and delicate as the flower, is a hidden beauty all its own.

Introduction to Superhydrophobicity

The lotus plant, part of the genus *Nelumbo*, is a water plant that is native to such parts of the world as Australia, Asia, and the Middle East. The stems of the lotus called rhizomes grow from the mud of marshes, lagoons, or ponds and branch off to its leaves and flowers. The rhizomes are anchored in the mud, supporting the great height of the lotus plant, which is capable of reaching 6 m. Large and disc-like, lotus

Particular gratitude to Caroline Hartel for allowing me to consult her draft on lotus leaves.

M. Lee (✉)
Mechanical Engineering, McCormick School of Engineering
at Northwestern University, Evanston, IL 60208, USA
e-mail: MichelleLee2013@u.northwestern.edu

Fig. 5.1 Lotus (*Nelumbo nucifera*) in Beijing (Photo by Professor Q. Wang)

leaves can grow up to 90 cm in diameter and can either float or hover above the water (Plant Cultures 2005).

The lotus leaf has been the subject of great interest in recent scientific research because of its superhydrophobicity, which is otherwise known as the Lotus Effect (Forbes 2008). Interestingly, the meaning of superhydrophobicity can be deduced from the components of the word: 'super-' which means 'very', '-hydro-' which means 'water', and '-phobicity' which is a variation of the word 'phobia'. Gathering these partial meanings together, it can be guessed that superhydrophobocity means something like 'very afraid of water'.

It turns out that this is precisely the sort of behavior that seems to characterize lotus leaves. When water hits the surface of a lotus leaf, it balls up into a bead as featured in Fig. 5.2 and rolls off, carrying dirt along with it (Quéré and Lafuma 2003). It is for this reason that lotus leaves are considered to be self-cleaning: despite being frequently surrounded by muddy water, lotus leaves are able to keep from being polluted and contaminated (Yan et al. 2011). This self-cleaning ability also explains why Buddhists saw the lotus plant as a symbol of purity (Solga et al. 2007).

The lotus leaf's superhydrophobic, self-cleaning surface is seen as an evolutionary advantage that benefits the life and longevity of the plant in a variety of ways. First, the existence of a water film on the plant surface has been proven to disturb the photosynthetic process by inhibiting gas exchange and increase the leaching of nutrients by a significant amount. Furthermore, the reduction of dust and

Fig. 5.2 A drop of water sitting on a lotus leaf exhibiting superhydrophobic properties (Photograph taken by Michael Gasperl, 2005)

contaminating particles is ideal, since dust can decrease photosynthesis by shading the plant and reducing diffusion of gases, among other negative impacts. Germination and penetration of pathogens such as fungi are also discouraged due to the dryness of the lotus leaf surface as well as its tendency to wash them away (Solga et al. 2007).

Though it is easy to gain a sense of what superhydrophobicity is based on the literal meaning of the word and the lotus leaf's behavior upon contact with water, there is much more to this phenomenon that can be attributed to various tribological factors. This chapter delves into the mechanisms at play on the leaf's surface that imparts its intriguing characteristics.

Wettability

Wettability and Contact Angle

The first contributing factor is wettability, which is directly correlated to the super-hydrophobicity of a surface. Wetting is defined as a liquid's ability to wet a surface, where a liquid is considered to have wet a surface if it spreads out completely (Zisman and Shafrin 1960). Wettability is determined by measuring the contact

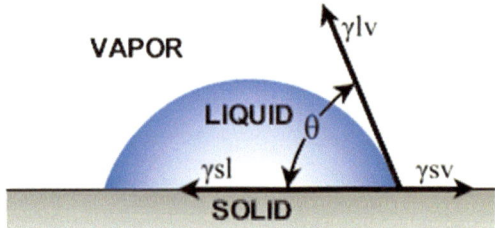

Fig. 5.3 Contact angle determines wettability and is governed by Young's Law. γ_{SL} is the interfacial tension of the solid-liquid interface, γ_{LV} is the interfacial tension of the liquid-vapor interface, and γ_{SV} is the interfacial tension of the solid-vapor interface (Courtesy of ramé-hart instrument company)

angle between the liquid and the surface. Contact angle, which is governed by Young's law, is found by solving:

$$\cos\theta_e = \frac{\gamma_{SV} - \gamma_{SL}}{\gamma_{LV}}$$

where θ_e is the equilibrium contact angle, γ_{SL} is the interfacial tension of the solid-liquid interface, γ_{LV} is the interfacial tension of the liquid-vapor interface, and γ_{SV} is the interfacial tension of the solid-vapor interface (McHale et al. 2011). See Fig. 5.3 for a visual representation.

Because a surface may not be uniformly flat, the contact angle may vary depending on the location on the surface. To take this variation caused by surface roughness or contamination into account, an equilibrium contact angle must be used (De Gennes 1985). Equilibrium contact angle depends on whether the liquid on the surface recedes or advances, and θ_e is defined as the difference:

$$\theta_e = \theta_A - \theta_R$$

between the advancing contact angle θ_A and receding contact angle θ_R (Agrawal 2012). Here, θ_A is measured upon an increase in contact area at the solid liquid interface, and θ_R is measured upon a decrease in contact area of said interface (De Gennes 1985).

Types of Wetting Surfaces

Generally, a wetting surface, or a hydrophilic surface, is characterized by a low contact angle, while a non-wetting surface, or a hydrophobic surface, is characterized by a high contact angle (Karthick and Ramesh 2008). In the case of superhydrophobic surfaces like the lotus leaf, equilibrium contact angles exceed 150°; in fact, the lotus leaf's equilibrium contact angle is found to be 170° (Roach 2003). The equilibrium contact angles of water for the other three types—superhydrophilic, hydrophilic, and hydrophobic—can be seen in Fig. 5.4.

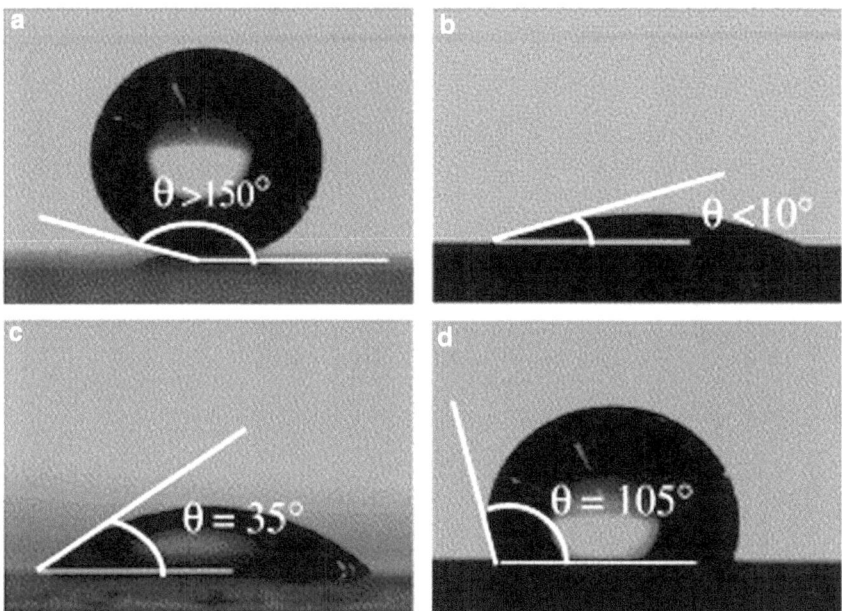

Fig. 5.4 Contact angles of water drop on a: (**a**) superhydrophobic surface fabricated by using microwave-plasma enhanced chemical vapor deposition (MPECVD), (**b**) superhydrophilic surface fabricated by using vacuum ultraviolet (VUV) light irradiation, (**c**) hydrophilic glass surface, and (**d**) hydrophobic glass surface coated with octadecyltrimethoxysilane (Reprinted from Wu et al. (2007). With permission from Elsevier)

The Role of Surface Roughness

The Cassie-Baxter (CB) Model

As stated previously, surface roughness is a key factor in the equilibrium contact angle. The lotus leaf's superhydrophobicity depends on the densities, diameters, and heights of wax tubules found on the upper side of the leaf that create a geometrically rough surface. According to the Cassie-Baxter (CB) model, which describes the wetting behavior of geometrically rough surfaces, the apparent contact angle, θ_e^*, can be defined as:

$$\cos\theta_E^* = r_\phi \phi_s \cos\theta_1 + \left(1 - \phi_x\right)\cos\theta_2$$

under the condition that the size of the liquid droplet is much greater than the size of the geometric features of the surface. In the CB model, r_ϕ is defined as the roughness or ratio of real contact area to the projected area of the region that is wetted, ϕ_s is the fraction of the area of the liquid-vapor interface obstructed by the texture, θ_1 is the equilibrium contact angle on solid phases, θ_2 is the equilibrium contact angle on air phases (=180°), and θ_e is the equilibrium contact angle (Choi et al. 2009).

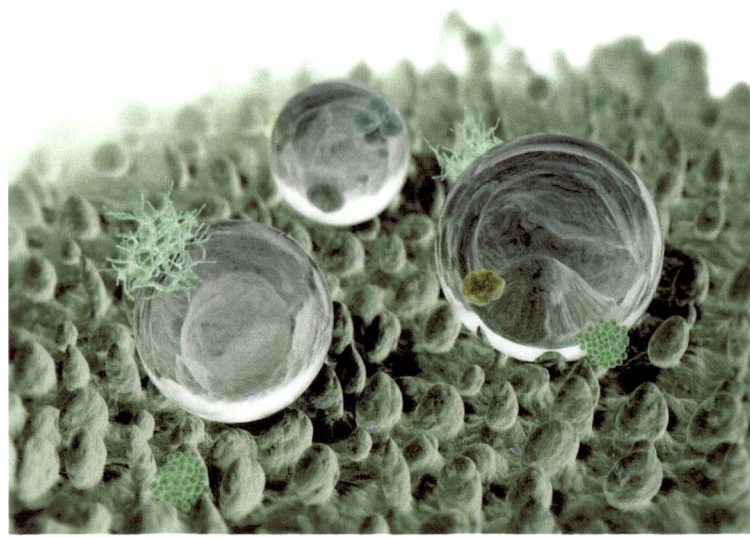

Fig. 5.5 Computerized representation of the microstructure of the lotus leaf's surface shows the form and variation of papillae, as well as dust particles picked up the water droplets as it sits on the superhydrophobic surface (Reprinted with kind permission from William Thielicke (w.th@gmx.de))

Hierarchical Structure of the Surface

Taking a closer look at the surface roughness of lotus leaves reveals why it plays such an important role in the leaves' superhydrophobicity. While the underside of lotus leaves are covered in convex cells and few stomata (Ensikat et al. 2011), papillae varying in shape and size populate the upper side of the lotus leaf (De Gennes 1985). The variation and structure of these papillae can be seen in the computerized representation of the surface's microstructure in Fig. 5.5. These papillae or micro-asperities are formed by papillose epidermal cells, and each of the papillae features its own nano-scale, tubule-like asperities made of epicuticular waxes. This structure, which features both micro- and nano-scale asperities, has come to be known as hierarchical structure (Bhushan and Jung 2011), shown at the far right of Fig. 5.6.

The hierarchical structure of the surface of lotus leaves has several characteristics that advance the surface's superhydrophobicity. First, the distinct shape of the microstructures, forming a micro-scale mountainous terrain with peaks and valleys, prevents water droplets from touching the entire surface. Because the valleys fill with air bubbles, water droplets sit on the apexes of the microstructures. This phenomenon happens on the nano-scale level as well due to the peaks and valleys formed by the papillae's nano-scale asperities, minimizing the contact area between the droplet and the surface (Bhushan and Jung 2011).

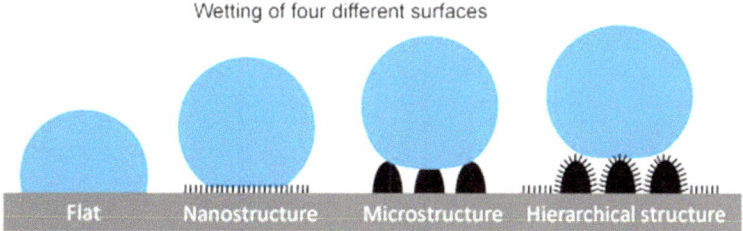

Fig. 5.6 Four differently structured surfaces elicit different wetting responses. The flat surface creates the largest area of contact between the drop and surface, with the microstructured surface next. The nanostructured surface provides less contact area than that of the microstructured surface, but the hierarchical structure generates the least contact area (Reprinted from Bhushan and Jung (2011). With permission from Elsevier)

Furthermore, though lotus leaves have the highest density of papillae population compared to other plant variety, their papillae are much smaller in diameter, which decreases the contact area between the surface and applied water droplets. The variation in the heights and diameters of the papillae also minimize contact area, as any water droplets on the surface will only rest on the highest peaks. An overview of the lotus leaf's surface can be seen in Fig. 5.7, which shows its key structures such as papillae and wax clusters and tubules, as well as the stomata and convex cells of the upper- and under-sides of the leaf, respectively (Ensikat et al. 2011).

Contact Area and Adhesion

The question we now have to ask is: why is minimized contact area so important? Ultimately, the minimization of contact area between a water droplet and the surface of a lotus leaf leads to a decrease in adhesion. In an experiment, researchers used atomic force microscopy (AFM) with a 15 μm radius tip to measure the adhesive forces on dried and fresh lotus leaves, among other plant variety such as colocasia and magnolia. They found the adhesive force to be greater for the fresh leaves in all instances. This was due to the fact that when the tip came into contact with the fresh leaf, the surface deformed because of its moisture content and pliability, creating a larger areal contact. In the case of the dry leaf, the lack of moisture prevented the surface from deforming, maintaining smaller areal contact. From this information, it can be concluded that a decrease in contact area in turn decreases adhesion (Bhushan and Jung 2011).

Adhesive forces also determine whether a water droplet will maintain its spherical shape or spread out upon contact with a surface. If the adhesive force from the surface is larger than the cohesive forces holding the sphere together, the droplet's shape will break and spread. Therefore, a surface characterized by a lower adhesive force such as the surface of a lotus leaf will allow a water droplet to maintain its shape, rather than spreading out and wetting the surface (Zisman and Shafrin 1960).

Fig. 5.7 (**a**) The upper surface of lotus leaves can repel water. (**b**) The hierarchical surface structure of the upper side of a lotus leaf is observed in a scanning electron microscopy (SEM) image, revealing papillae, wax clusters, and wax tubules. Leaf was prepared utilizing 'glycerol substitution'. (**c**) Wax tubules. (**d**) Wax tubules dissolved following critical-point drying. (**e**) Convex cells on the underside of a leaf following critical-point drying (Reproduced with kind permission from Ensikat et al. 2011)

Engineering Applications

Self-Cleaning Windows

The wonders of the Lotus Effect have been taken advantage of in a wide variety of markets, ranging from self-cleaning windows to stain-resistant clothing. In fact, the term 'Lotus Effect' was coined by botanist Wilhelm Barthlott, who first predicted that a synthetic material mimicking the waxy, rough surface of the lotus leaf could be very useful in engineering applications. He registered the Lotus Effect as a trademark when he patented the idea to produce self-cleaning surfaces with raised microstructures. Since then, self-cleaning windows inspired by Barthlott's idea have been made. Just as dirt is dissolved and rolled away by water droplets on the surface of the lotus leaf, windows have been synthetically modified to be superhydrophobic so that dirt can roll right off. This kind of modification eliminates the need for window-washers, as rainwater would do the cleaning (Forbes 2008).

Building Protection

Similarly, self-cleaning paint known as Lotusan® has been made by Sto Corp. utilizing the Lotus Effect, featuring an exterior coating called StoCoat® that can also be applied to the exteriors of buildings to achieve the self-cleaning effect. StoCoat® was created to be extremely water-repellent by minimizing the contact area for dirt and water using models of the micro- and nano-structured topography of lotus leaves (Sto Corp. 2004).

The benefits of applying the lotus leaf's superhydrophobic, self-cleaning ability to buildings go beyond aesthetic improvements. Various researchers have shown the ability of numerous fungi, algae, and bacteria to attach to and colonize building structures, affecting concrete, roof tiles, facades, and more. Currently, biocides are often used to remove such colonies, as they are known to cause biodeterioration costing an estimated 2–4 billion € annually in Germany alone. However, the toxicity of such biocides makes the environmentally safe lotus effect an attractive alternative.

In an effort to determine the effectiveness of superhydrophobic, self-cleaning paint compared to conventional paints, Solga et al. exposed six test specimens under deciduous trees for 6 years. Four specimens were of conventional paints, while the other two were of the self-cleaning paint. The results showed vast differences between specimens of the two paints; while the four conventionally-painted samples were largely covered in green algae, the two remaining samples were clean except for the very bottoms, where they were sometimes flooded with rain. Based on these results, researchers concluded that there indeed existed better alternatives to harmful biocides in preventing biodeterioration of buildings (Solga et al. 2007).

More Applications

Stain-resistant clothing has also been developed based on the Lotus Effect. Called Aquapel®, this fabric manufactured by Nano-Tex® is specially treated by being modified at the molecular level with hydrophobic 'whiskers' that attach to individual fibers of the fabric. These extra whiskers elevate drops of liquid much like the lotus leaf and cause them to bead up and roll off (Nano-Tex 2013).

Additionally, potential applications for the Lotus Effect include coatings on radio car antennas to keep them free of snow and ice, as well as coatings on hulls of ships to allow them to traverse waters more easily (Roach 2003). If applied to airplane wings, this would also eliminate the need for frequent de-icing.

As is evident in this chapter, despite being the seemingly humble counterpart to the elegant lotus flower, the lotus leaf boasts many, many remarkable characteristics that engineers and scientists have been inspired by and have tried to mimic. Its superhydrophobicity and hierarchical surface structure has fascinated researchers, and some innovations based on these characteristics have already proven successful in certain applications. The lotus leaf is just one of countless botanical surfaces that bioinspire, and others will be introduced later in this book.

References

Agrawal A (2012) Wettability, non-wettability and contact angle hysteresis. http://web.mit.edu/nnf/education/wettability/wetting.html. Accessed 27 Nov 2012

Bhushan B, Jung YC (2011) Natural and biomimetic artificial surfaces for superhydrophobicity, self-cleaning, low adhesion, and drag reduction. Prog Mater Sci 56(1):1–108. doi:10.1016/j.pmatsci.2010.04.003

Choi W, Tuteja A, Mabry JM, Cohen RE, McKinley GH (2009) A modified Cassie-Baxter relationship to explain contact angle hysteresis and anisotropy on non-wetting textured surfaces. J Colloid Interface Sci 339(1):208–216. doi:10.1016/j.jcis.2009.07.027

De Gennes PG (1985) Wetting: statics and dynamics. Rev Mod Phys 57(3):832–836. doi:10.1103/RevModPhys.57.827

Ensikat HJ, Ditsche-Kuru P, Neinhuis C, Barthlott W (2011) Superhydrophobicity in perfection: the outstanding properties of the lotus leaf. Beilstein J Nanotechnol 2:152–161. doi:10.3762/bjnano.2.19

Forbes P (2008) Self-cleaning materials: lotus leaf-inspired nanotechnology. http://insurftech.com/docs/links/Related-Papers/Article-1-Scientific-American-Self-Cleaning-Materals-Lotus-Effect.pdf. Accessed 16 July 2013

Karthick B, Ramesh M (2008) Lotus-inspired nanotechnology applications. Resonance 13(12):1141–1145. doi:10.1007/s12045-008-0113-y

McHale G, Newton M, Shirtcliffe NJ, Geraldi NR (2011) Capillary origami: superhydrophobic ribbon surfaces and liquid marbles. Beilstein J Nanotechnol 2:145–151. doi:10.3762/bjnano.2.18

Nano-tex (2013) Aquapel. http://www.nanotex.com/technologies/aquapel.html. Accessed 27 Nov 2012

Plant Cultures: Exploring plants & people (2005) Lotus—plant profile. http://www.kew.org/plant-cultures/plants/lotus_plant_profile.html. Accessed 16 July 2013

Quéré D, Lafuma A (2003) Superhydrophobic states. Nat Mater 2:457–460. doi:10.1038/nmat924

Roach J (2003) New water-repellent material mimics lotus leaves. http://news.nationalgeographic.com/news/2003/02/0227_030227_lotusmaterial.html. Accessed 16 July 2013

Solga A, Cerman Z, Striffler BF, Spaeth M, Barthlott W (2007) The dream of staying clean: lotus and biomimetic surfaces. Bioinsp Biomim 2:S126–S134. doi:10.1088/1748-3182/2/4/S02

Sto Corp. (2004) StoCoat Lotusan. http://www.stocorp.com/index.php/component/option,com_catalog2/Itemid,196/catID,43/catLevel,5/lang,en/productID,34/subCatID,44/subCatIDBP,44/subCatIDnext,0/. Accessed 27 Nov 2012

The Flower Expert (2005) Lotus flowers. http://www.theflowerexpert.com/content/aboutflowers/exoticflowers/lotus. Accessed 16 July 2013

Wu Y, Kouno M, Saito N, Nae FA, Inoue Y, Takai O (2007) Patterned hydrophobic-hydrophilic templates made from microwave-plasma enhanced chemical vapor deposited thin films. Thin Solid Films 515(9):4203–4208

Yan YY, Gao N, Barthlott W (2011) Mimicking natural superhydrophobic surfaces and grasping the wetting process: a review on recent progress in preparing superhydrophobic surfaces. Adv Coll Int Sci 169(2):80–105. doi:10.1016/j.cis.2011.08.005

Zisman WA, Shafrin EG (1960) Constitutive relations in the wetting of low energy surfaces and the theory of the retraction method of preparing monolayers. J Phys Chem 64(5):519–524. doi:10.1021/j100834a002

Dragonfly Wings: Special Structures for Aerial Acrobatics

Michelle Lee

Ancient Aerial Acrobats

The crowd watches with bated breath as the performers maneuver their way through and around the hoop, bending in impossible ways. Many feet above the ground, the acrobats defy gravity as they assume forms so different from anything remotely imaginable, accomplishing it with grace and balance all the while. Except for the occasional gasp, not a sound is heard from the audience for fear that a word might shatter the delicateness of the act on stage. Their movements are so deliberate and beautiful, and it is only after their feet touch the ground again that the tension in the air is broken by erupting applause, echoing throughout the theater (Fig. 6.1).

As people file out of the theater's grand doors after the show, many think they will never see anything like it again, that it is truly a one-of-a-kind wonder. Despite the magnificence of the acrobatic show, however, there is yet another kind of aerial acrobat that deserves its own show: the naturally gifted, aerodynamic dragonfly. In fact, dragonflies are the original aerial acrobats, having existed on Earth for more than 250 million years (Griffith 2006).

Over history, dragonflies (Fig. 6.2) have been found across the globe, allowing a rich multitude of culture and symbolism to be developed around these four-winged creatures. For example, in Native American history, dragonflies were symbols of activity and swiftness and were often associated with horses. Japanese history surrounding dragonflies may be the most interesting, however, as these insects were considered to serve as winged mounts for the Hotoke-Sama, or August Spirits of the Ancestors. Among Buddhists, the Hotoke-Sama were thought to return on August 15th,

Particular gratitude to Christopher Timpone for allowing me to consult his draft on dragonfly wings.

M. Lee (✉)
Mechanical Engineering, McCormick School of Engineering
at Northwestern University, Evanston, IL 60208, USA
e-mail: MichelleLee2013@u.northwestern.edu

M. Lee (ed.), *Remarkable Natural Material Surfaces and Their Engineering Potential*,
DOI 10.1007/978-3-319-03125-5_6, © Springer International Publishing Switzerland 2014

Fig. 6.1 Aerial acrobats performing in Misawa City, Japan during the Cirque Dreams Jungle Fantasy World Tour on January 18th (Photo taken by Tech. Sgt. Marie Brown for the U.S. Air Force)

Fig. 6.2 *Crocothemis erythraea* on a leaf (Photo by Ejatgc)

riding dragonflies into their old homes to be reunited with their families (Mitchell and Lasswell 2005).

Though such folklore may have diffused a bit over time as scientific research dedicated to dragonflies began, fascination with their flying and maneuvering capabilities has not, perhaps even increasing in recent years. One look at the dragonfly's impressive flying abilities can convince that this attention is well afforded: they can fly sideways, forwards and backwards, hover in midair and reverse directions instantaneously, accelerate rapidly, and fly as fast as 50 km/h (Rajabi et al. 2011). Although dragonfly wings account for less than 2 % of the total body mass, they are the main enablers of such diverse flight behavior (Sun and Bhushan 2012). This chapter discusses the characteristics of dragonfly wings that make them so remarkable for flight, as well as their many engineering applications. Perhaps aerial acrobats will even learn a thing or two for their next show.

Membrane

Purpose and Composition

During flight, the membranes of dragonfly wings experience many different aerodynamic forces, requiring them to take a variety of shapes and positions. For example, when wings twist, lift production can be increased for further upward force; in addition, thrust production can be enhanced with certain deformations of the membranes (Song et al. 2007). When all four wings beat together, they can experience spanwise torsional deformation and coupled bending along the length of the dragonfly (Ren et al. 2012).

To understand the aerodynamic and mechanical properties of dragonfly wings, it is necessary to be acquainted with their basic structure. The first main attribute of dragonfly wings is its thin, transparent, film-like membrane, which is cuticular and supported by a framework of veins (Sun and Bhushan 2012). The membrane is only a few micrometers thick and is composed mainly of chitin, a nitrogenous polysaccharide that is often found in the shells of shrimps and crabs, as well as in the husks or shells of insects. Composed of a structure similar to plant cellulose, chitin is very elastic and flexible (Yoshihara et al. 2012). The chitin is embedded in a matrix in fibrillar form containing multiple structural lipids and proteins (Kreuz et al. 2001).

In their microstructure tests of the dragonfly wing membrane, Song et al. discovered that the membrane was not singular and uniform, but it was in fact separated into three layers: the ventral, middle, and dorsal layers. These layers, which can be seen in Fig. 6.3a, were measured with Image Analysis System (IAS) and found to be 356.33 ± 42.50 nm, 1.93 ± 0.18 μm, and 513.63 ± 69.02 nm, respectively. The dorsal and ventral layers were revealed to be covered in nano-scale columnar wax structures that were almost vertically erected, as shown in Fig. 6.3b. The wax columns were not arranged in an orderly fashion, and each column had a round cap

a

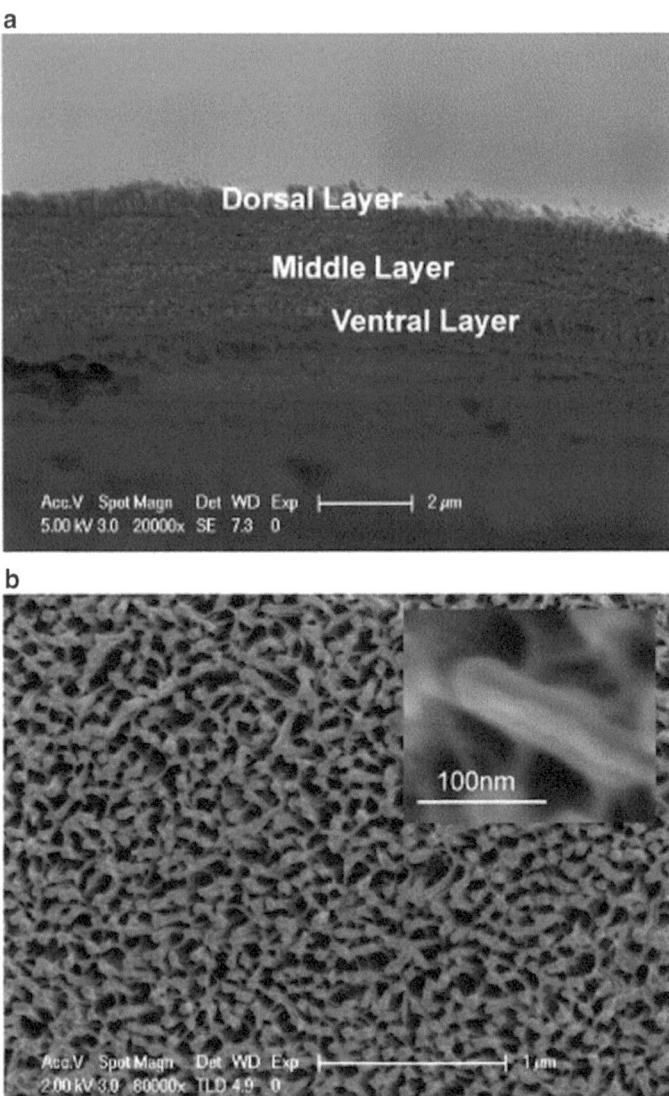

b

Fig. 6.3 Scanning electron microscopy images show: (**a**) the three layers of the cross-section of the wing membrane, namely the dorsal, middle, and ventral layer, and (**b**) the surface of the cuticle waxy layer with wax columns characterized by round caps and nanogrooves (*inset*) (Reprinted from Song et al. (2007). With permission from Elsevier)

resembling a hemisphere and grooves along its length (Fig. 6.3b, inset). These nano-scale wax features of the wing membrane were found to be superhydrophobic with a contact angle of about 174°, leading Song et al. to conclude that the superhy-drophobicity aids in flight during the rain and enables the wings to self-clean (Song et al. 2007).

Surface Corrugation

The membrane of a dragonfly wing is not smooth, and when the cross section is examined, a very well-defined corrugation can be seen. The almost-jagged, corrugated configuration is of utmost importance in stabilizing the thin, light wing (Kesel 2000). In order to analyze the corrugation of dragonfly wings, Jongerius and Lentink scanned the forewing and hindwing of *Sympetrum vulgatum* with a micro computed tomography (micro-CT) scanner, which uses X-ray imaging to stack 3D cross-sections of the wing. The researchers wrote a Matlab code to smooth and connect the 3,997 cross-sectional images of the forewing produced by the scanner to generate a digital reconstruction of the wing. Figure 6.4 shows the forewing and its corrugation profile (Jongerius and Lentink 2010).

Based on a 2-dimensional wing model used by Kesel in previous experiments, Kim et al. investigated wing corrugation's effect on lift and drag. For corrugated models with different profiles 1, 2, and 3, modified versions in which the corrugations were filled with straight lines were examined. These modified versions thus represented uncorrugated surfaces and were labeled profiles 1A, 2A, and 3A (Kim et al. 2009).

Results from a numerical analysis of profiles 1 and 1A at various Reynolds numbers revealed that the corrugated profile (profile 1) corresponded to an increase in lift coefficients for all cases. In Fig. 6.5, the effect of Reynolds number on lift and

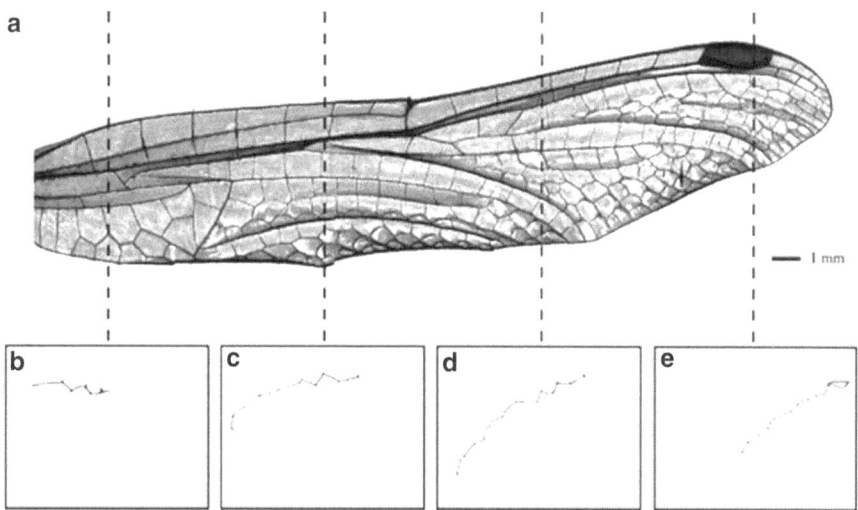

Fig. 6.4 Digital reconstruction using micro computed tomography (micro-CT) scanner shows: (**a**) forewing of *Sympetrum vulgatum*, with thin areas lighter and thick areas darker (linear progression from light to dark), and (**b–e**) corrugation profiles of the cross-section of the forewing. Note greater corrugation near the root of the wing, towards (**b**) (Reprinted from Jongerius and Lentink (2010), Fig. 3. With kind permission from Springer Science + Business Media)

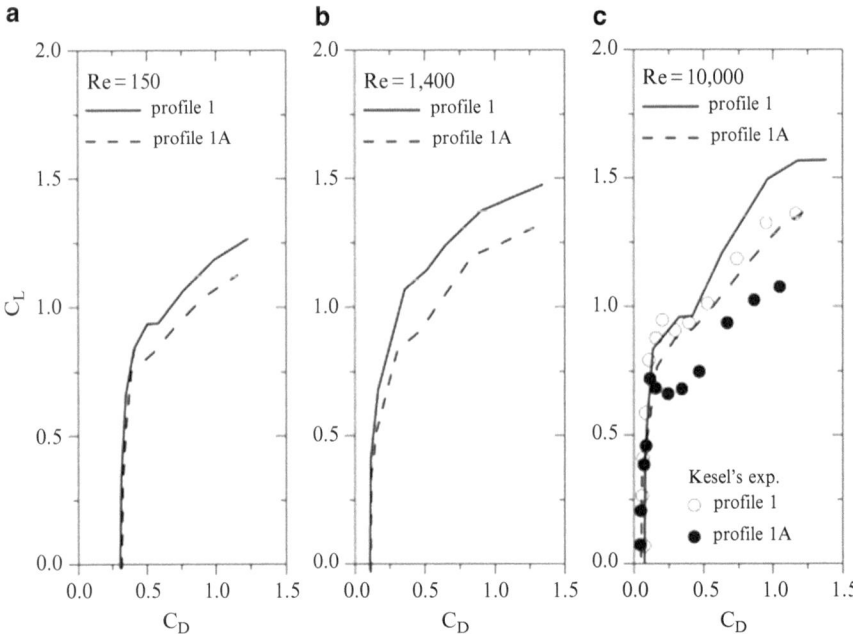

Fig. 6.5 (a–c) Polar diagram of corrugated profile 1 and smooth profile 1A at varying Reynolds numbers (Reprinted from Kim et al. (2009). With permission from Elsevier)

drag coefficients can be seen, as well as the fact that the lift coefficients for the corrugated profile was always higher (Kim et al. 2009). Therefore, at low Reynolds numbers, corrugation helps produce enhanced aerodynamic performance (Jongerius and Lentink 2010).

Corrugation also increases the strength and stiffness of the wing and its ability to absorb stress against bending in the spanwise direction. It additionally increases flexural rigidity, helping to prevent fracture due to fatigue (Sun and Bhushan 2012). According to Newman and Wootton, wing corrugation plays a critical role in the dragonfly's response to excessive loading (Ren et al. 2012), such as those inflicted upon the wings due to collisions or encounters with particularly sudden or strong gusts of wind. They identify two possible adaptive strategies—incredible stiffness and strength enabling wings to withstand any load or brief yielding followed by full recovery—and point out that dragonflies employ the second, preferred strategy. While the first strategy would demand wing characteristics that might negatively impact aerodynamic performance (e.g. deep corrugation), the second renders dragonfly wings flexible and almost unbreakable, as it would prevent any one component of the wing from being loaded beyond capacity (Newman and Wootton 1986).

Veins

Primary Hierarchical Level: Sandwich Structure

Hierarchy is ubiquitous in nature. Many natural and biological materials—shell, teeth, bone, and mother of pearl, to name a few—have evolved to consist of multiple levels of hierarchy, enhancing their biomechanical properties. The veins in dragon-fly wings are no exception, and their hierarchical structure lends them to superior mechanical properties (Chen et al. 2012). The framework of veins supporting the membrane is critical to the performance of the dragonfly's wings. In fact, the strength of dragonfly wings depends mostly on the structure of the veins (Wang et al. 2008). This section will delve into the important properties of the veins and how they affect dragonfly flight.

The veins, which are hollow and tubular, branch out and criss-cross, forming cells that range from rectangular shapes at the leading edge to various polygons at the trailing edge (Sun and Bhushan 2012). Found to have a complex sandwich like microstructure, these veins have an inner and outer shell made of chitin and a mid-dle layer of protein and muscle, which can be seen in Fig. 6.6. The chitinous shell is macroscopically brittle, while the muscle and protein layer is microscopically duc-tile, creating a mixed fracture model that, according to Wang et al., may be respon-sible for the ability of the wings to stiffen against aerodynamic bending moments, as well as torsional deformation (Wang et al. 2008).

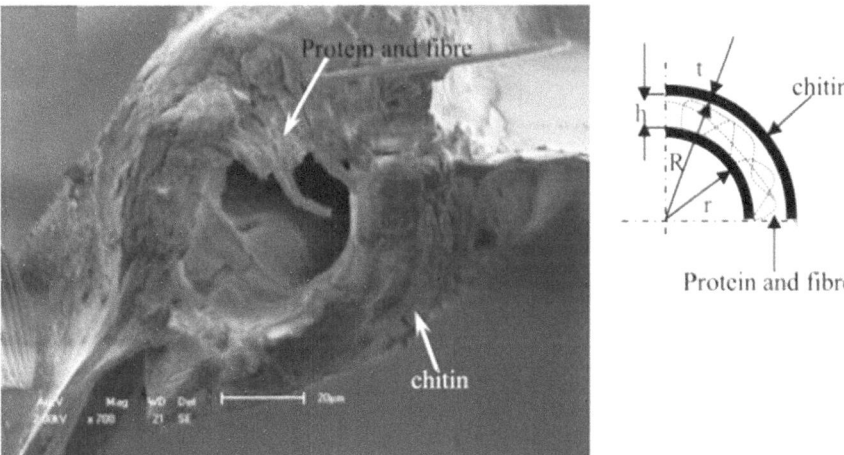

Fig. 6.6 Sandwich microstructure of a tubular vein (Reprinted from Wang et al. (2008). With permission from Elsevier)

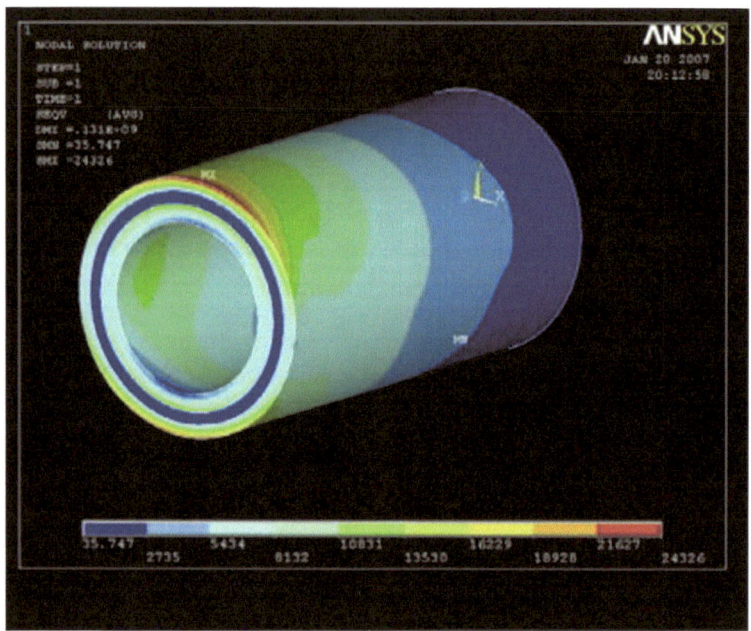

Fig. 6.7 Finite element analysis (FEA) of constrained end shows von Mises stress (Reprinted from Wang et al. (2008). With permission from Elsevier)

The composite sandwich structure of these veins proves superior to those of a single material, because the different layers are responsible for receiving different loads during flight. The muscle and protein layer receives the torsional deformation, while the chitinous shell layer receives the bending load. In this model, neither of the layers is easily damaged. For example, the chitinous shell layer is not easily damaged when the wing must experience torsional deformation, because the sandwiched layer takes the strain upon itself.

Wang et al. validated this behavior through finite element analysis (FEA), as shown in Figs. 6.7 and 6.8. When the model was subject to bending loads in Fig. 6.7, the greatest stress was evident on the outermost layer and at the constrained end, and the chitin layers bore the bulk of the load. The middle layer of the sandwich structure was shown to experience the least stress. Figure 6.8 shows the most torsional deformation to be at the interface between the sandwiched protein and the chitin shell when the vein undergoes torsional moment (Wang et al. 2008).

In another experiment, Chen et al. found by subjecting a computer model to a bending load that the shell layer bears 30 % larger Von Mises stress with the soft protein layer than without. The interface between the protein and shell layer contributes even more to the bearability of large torsional loads, because it allows some rotation between the two (Chen et al. 2012).

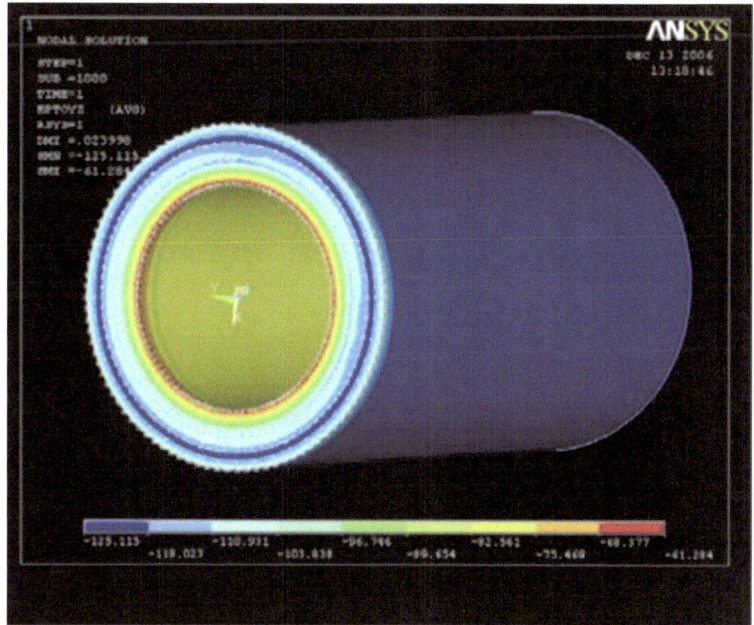

Fig. 6.8 Finite element analysis (FEA) of free end shows strain in Y-Z direction (Reprinted from Wang et al. (2008). With permission from Elsevier)

Secondary Hierarchical Level: Multilayered Chitinous Shell and Protein Fibrils

The secondary hierarchical level of the vein features two major properties, the first being the multilayered structure of the outer and inner chitinous shells. These shells consist of concentric layers of chitin, with thicknesses ranging from 600 to 700 nm. The slip between these layers allows the shell to afford larger deformation under loading (Chen et al. 2012).

The other property of the secondary hierarchical level is the growth of fibrils in the protein layer. The fibrils, whose diameters are no more than 100 nm, strengthen and toughen the soft matter and grow in clusters on the order of several micrometers in size. They grow along the circumferential direction, which is critical to the torsional load bearing capacity of the protein layer, because it is mainly along the circumference that the protein layer deforms when the wing warps. As they grow, the clusters become taut in favorable ways that absorb torsion deformation, relieving the overall vein structure from demands of the load (Chen et al. 2012). Figure 6.9 displays an overview of the primary and secondary hierarchical levels.

The combination of vein framework and elasticity of the membrane is the basis for the mechanical strength that dragonfly wings have that have allowed them to develop such high flying abilities (Yoshihara et al. 2012).

Fig. 6.9 Hierarchical vein
microstructure includes:
(**a**) composite sandwich
structure comprised of
chitinous shell with protein
layer, (**b**) concentric layers
about 600–700 nm thick
within the chitinous shell,
and (**c**) circumferentially-
growing fibrils in the protein
layer (Reprinted from Chen
et al. (2012). With permission
from Elsevier)

a

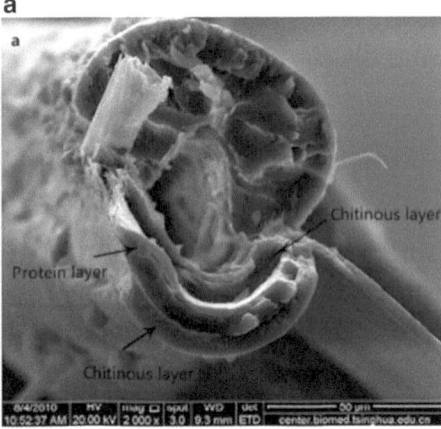

b

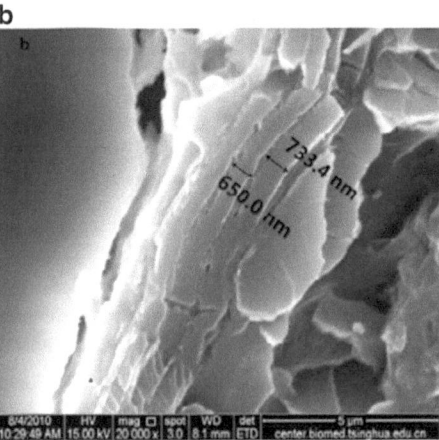

c

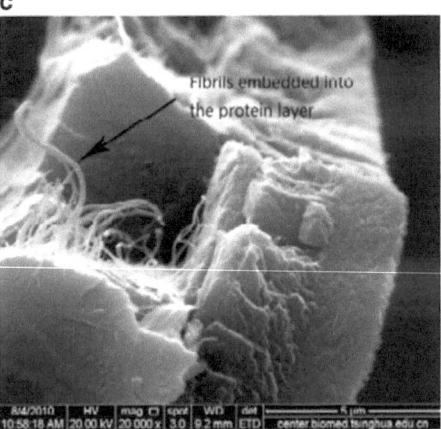

Micro- and Nano-scale Ripple Morphologies

Another property of dragonfly wing veins that is worth examining is the ripple-like surface texture. Based on the principle that dragonflies and other small-scale flying devices cannot neglect micro- and nano-scale surface structures (as opposed to large-scale ones like airplanes, that can ignore the effects of such relatively insignificant structures), Zhao et al. took a closer look at the surface morphologies of dragonfly wings, specifically the *Pantala flavesens* Fabricius, through Environmental Scanning Electron Microscope (ESEM). Subsequently, the ESEM images revealed micro- and nano-scale ripple morphologies along the surfaces of the veins (Fig. 6.10).

Upon finding that the *Crocothemis servilia* Drury dragonfly exhibited the same ripple texture on its wings, Zhao et al. concluded that this was a universal property of dragonflies and likened the lift effect of these ripples to the pitted face of a golf ball (Zhao et al. 2010). Pitted golf balls can travel nearly twice as far as smooth ones, because it forces the boundary layer of air to become turbulent, narrowing the wake of low-pressure air. With a smaller wake, the pressure difference between the front and back of the ball is less, reducing the pressure drag (Shelton 2007). According to Zhao et al., a similar mechanism may be at play during a dragonfly's flight due to the ripple waves on the veins (Zhao et al. 2010).

Vein-Joints

As previously mentioned, wing flexibility is important in the dragonfly's ability to recover from large loads. In Donoughe et al.'s observations of the winged order Odonata, which includes dragonflies and damselflies, vein-joints—the junctions at

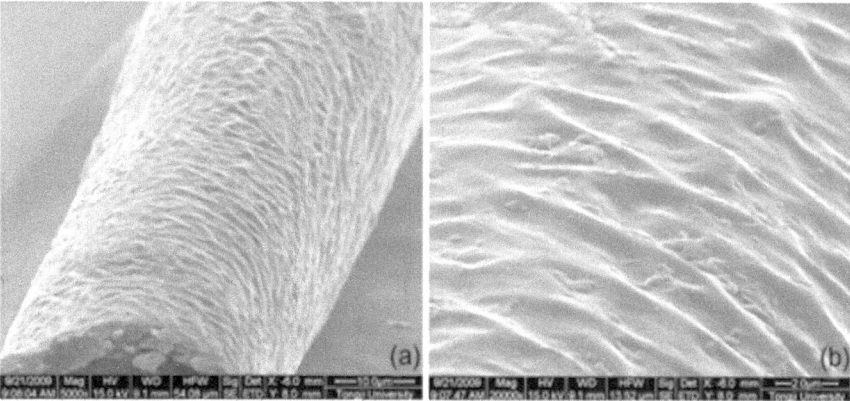

Fig. 6.10 (**a**) A longitudinal vein's ripple-like texture and (**b**) magnified ripple-like texture (Reprinted from Zhao et al. (2010), Fig. 2. With kind permission from Springer Science+Business Media)

which longitudinal veins and cross-veins meet—were found to be key contributors to wing flexibility. Dragonflies' vein-joints were discovered by Gorb in 1999 to contain resilin, a protein present in various insects, such as in the legs of fleas and the bases of locust wings. Resilin is known for its great flexibility, with its Young's modulus at about 1 MPa, and its superb energy-storing capabilities. In the vein-joints containing resilin, the protein was found to act like a flexible hinge, providing significant chordwise flexibility; this agrees with the discovery by Combes and Daniel in 2003, in which chordwise flexural stiffness was found to be up to two orders of magnitude smaller than that of the spanwise direction (Donoughe et al. 2011).

Engineering Applications

Micro-Air-Vehicles

Micro-Air-Vehicles, or MAVs, are miniature, palm sized devices that have been met with much enthusiasm by the Defense Advanced Research Agency (DARPA) and the aerospace industry for information gathering. MAVs may be equipped with sur-veillance technology like optical, sensory, control, and communication devices and are being developed to fly into and monitor hazardous or remote environments (Tamai 2007). Another area in which the utilization of MAVs is being considered is homeland security (Sun and Bhushan 2012).

DARPA and the aerospace industry have high hopes for the little fliers, but there are still many aspects in which MAVs can improve, especially if it is to fulfill certain expectations. These include being able to maintain hovering and forward flight, fly-ing in tunnels, caves, and urban settings, and maneuvering in confined spaces. To address the areas in which MAVs are lacking, such as being able to accommodate variations in wind gusts and maintaining flight stability, engineers and researchers are looking to the top experts in flight: birds and insects (Shyy et al. 2008).

However, a brief comparison of the bird, hummingbird, butterfly, and dragonfly reveals that the dragonfly is the optimal design choice for MAV modeling. When considering other animals, issues arise such as having too complex muscle coordi-nations (bird), lack of power efficiency (hummingbird), and low agility (butterfly). The dragonfly, on the other hand, boasts many positive attributes, including unparal-leled maneuverability, simple control, extremely long flight times, and resistance to damage. As such, the dragonfly is an excellent model for MAVs, and novel designs have already been developed based on research of the dragonfly (Ratti and Vachtsevanos 2012).

Despite accounting for such a tiny percentage of the dragonfly's weight, dragon-fly wings are complex structures with hierarchical structure, micro- and nano-scale surface morphologies, and many more unique and interesting features that consis-tently intrigue researchers. They are loaded with biomimetic potential and have already proven useful in engineering applications, particularly in the design of micro-air-vehicles. As scientists continue to explore the structural and topographi-cal mechanisms of dragonflies, nature's ingenuity and brilliance will be revealed yet again and again in these aerial acrobats.

References

Chen Y, Wang X, Ren H, Yin H, Jia S (2012) Hierarchical dragonfly wing: microstructure-biomechanical behavior relations. J Bionic Eng 9:185–191. doi:10.1016/S1672-6529(11)60114-5

Donoughe S, Crall JD, Merz RA, Combes SA (2011) Resilin in dragonfly and damselfly wings and its implications for wing flexibility. J Morphol 272(12):1409–1421. doi:10.1002/jmor.10992

Griffith V (2006) Aerial acrobats: from the Ozarks to Africa, dragonflies take flight through entomologist's field study, photographs and interactive site. http://www.utexas.edu/features/2006/dragonflies/index.html. Accessed 26 Aug 2013

Jongerius SR, Lentink D (2010) Structural analysis of a dragonfly wing. Exp Mech 50:1323–1334. doi:10.1007/s11340-010-9411-x

Kesel AB (2000) Aerodynamic characteristics of dragonfly wing sections compared with technical aerofoils. J Exp Biol 203:3125–3135

Kim WK, Ko JH, Park HC, Byun DY (2009) Effects of corrugation of the dragonfly wing on gliding performance. J Theor Biol 260(4):523–530. doi:10.1016/j.jtbi.2009.07.015

Kreuz P, Arnold W, Kesel AB (2001) Acoustic microscopic analysis of the biological structure of insect wing membranes with emphasis on their waxy surface. Ann Biomed Eng 29(12):1054–10581. doi:10.1114/1.1424921

Mitchell FL, Lasswell JL (2005) A dazzle of dragonflies. Texas A&M University Press, College Station

Newman DJS, Wootton RJ (1986) An approach to the mechanics of pleating in dragonfly wings. J Exp Biol 125:361–372

Rajabi H, Moghadami M, Darvizeh A (2011) Investigation of microstructure, natural frequencies and vibration modes of dragonfly wing. J Bionic Eng 8:165–173. doi:10.1016/S1672-6529(11)60014-0

Ratti J, Vachtsevanos G (2012) Inventing a biologically inspired, energy efficient Micro Aerial Vehicle. J Intell Robot Syst 65:437–455. doi:10.1007/s10846-011-9615-z

Ren HH, Wang XS, Chen YL, Li XD (2012) Biomechanical behaviors of dragonfly wing: relationship between configuration and deformation. Chin Phys B 21(3):03501. doi:10.1088/1674-1056/21/3/034501

Shelton M (2007) Probing question: how do dimples make golf balls travel farther? http://news.psu.edu/story/141235/2007/06/18/research/probing-question-how-do-dimples-make-golf-balls-travel-farther. Accessed 25 July 2013

Shyy W, Lian Y, Tang J, Liu H, Trizila P, Stanford B, Bernal L, Cesnik C, Friedmann P, Ifju P (2008) Computational aerodynamics of low Reynolds number plunging, pitching and flexible wings for MAV applications. Acta Mech Sin 24:351–373. doi:10.1007/s10409-008-0164-z

Song F, Xiao KW, Bai K, Bai YL (2007) Microstructure and nanomechanical properties of the wing membrane of dragonfly. Mat Sci Eng A 457(1–2):254–260. doi:10.1016/j.msea.2007.01.136

Sun J, Bhushan B (2012) The structure and mechanical properties of dragonfly wings and their role on flyability. CR Mecanique 340:3–17. doi:10.1016/j.crme.2011.11.003

Tamai M (2007) Experimental investigations on biologically inspired airfoils for MAV applications. Thesis, Iowa State University

Wang XS, Li Y, Shi YF (2008) Effects of sandwich microstructures on mechanical behaviors of dragonfly wing vein. Compos Sci Technol 68(1):186–192. doi:10.1016/j.compscitech.2007.05.023

Yoshihara A, Miyazaki A, Maeda T, Imai Y, Itoh T (2012) Spectroscopic characterization of dragonfly wings common in Japan. Vib Spectrosc 61:85–93. doi:10.1016/j.vibspec.2012.03.010

Zhao HX, Yin YJ, Zhong Z (2010) Micro and nano structures and morphologies on the wing veins of dragonflies. Chin Sci Bull 55(19):1993–1995. doi:10.1007/s11434-010-3253-x

Moth Eyes: A New Vision for Light-Harnessing Efficiency

Michelle Lee

Award-Winning Theatrical Actors

Everyone's head turns as a slick black limousine pulls up to the red carpet, and before the door even opens, light bulbs from cameras flash left and right with no sign of ceasing. The noise is relentless: journalists, photographers, and fans shout and scream, trying to get a better look at the emerging pair of celebrities. Jaws hang open in awe of her shiny, brown hair, falling in light waves around her shoulders. Her dress is a shimmery gold, draping down to the floor like liquid metal, and her lips are bright and bold, the color matching that of the red carpet. Her arm is hooked around the arm of a man with a tailored black suit and perfectly styled hair. His shoes are so shiny that they reflect the flashes coming from the cameras, which are all trained on them. Together, they drift up the red carpet, the sound of their names and of their latest hit movie following after them like the gold train of her dress.

Actors and actresses are some of the most famous and loved people in America. They are surrounded in glamour, beauty, and fame, and experiences like the above are a given. If the insect world had celebrities, surely they would be the flashy, iridescent butterfly or the delicately spotted ladybug. Perhaps spiders would make the cut as well, as they could play the terrifying villains. Moths, on the other hand, seem to go rather unnoticed compared to other types of insects. In the event that they do get noticed, it is usually followed up by a trip to the store for mothballs. However drab, pesky, and relatively average moths may seem compared to some of their flashy friends, these little furry-winged insects have their own secret talent.

Just like humans look to methods of biomimicry to solve real world problems, it turns out moths know a thing or two about mimicry as well. As a means of survival, *Brenthia* moths thwart jumping spider predators by mimicking them, as in Fig. 7.1. When encountering jumping spiders, these particular moths survive more often than

M. Lee (✉)
Mechanical Engineering, McCormick School of Engineering
at Northwestern University, Evanston, IL 60208, USA
e-mail: MichelleLee2013@u.northwestern.edu

M. Lee (ed.), *Remarkable Natural Material Surfaces and Their Engineering Potential,*   79
DOI 10.1007/978-3-319-03125-5_7, © Springer International Publishing Switzerland 2014

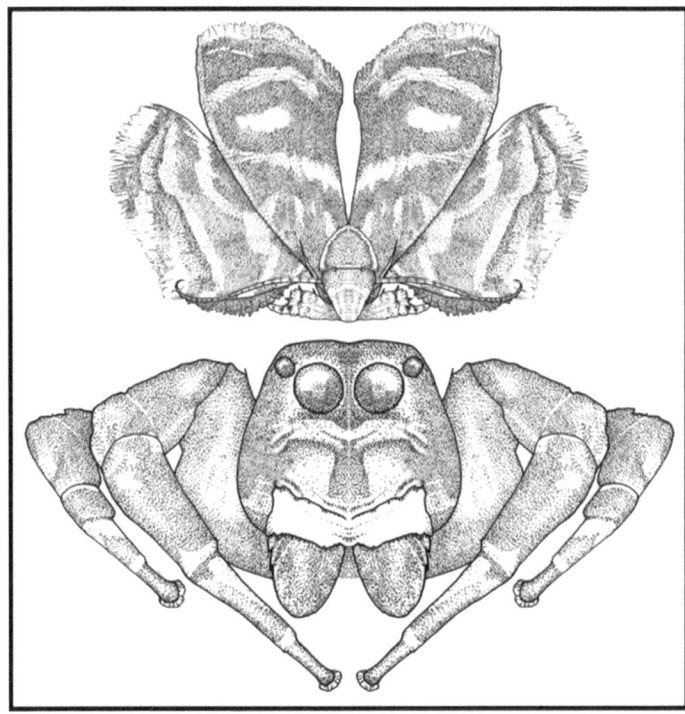

Fig. 7.1 *Brenthia* moths mimic spiders for survival (Drawing by Virginia Wagner. Reproduced with kind permission from Rota and Wagner (2006))

controls. The territorial behavior of jumping spiders upon meeting a *Brenthia* moth while it is mimicking is further evidence that the spiders recognize the moths as their own kind (Rota and Wagner 2006).

Amazingly, moths are capable of mimicking predators that are much larger than them, such as foxes and snakes. According to retired entomologist Professor Philip Howse, moths can fool snakes with the patterns on its wings; in an effort to fool a predator even further, the mimicking moths will even fall and writhe on the ground, adding the perfect touch to their theatrical acts. Other species use the patterns on their wings to look like foxes and rodents peeking out from greenery. The reason this works so well with animals and not humans is that animals have more focused vision, and they tend to train their eyes on small details (Gray 2010).

The ability to mimic a predator for survival is a talent that not many animals have—it is considered both rare and exceptional (Rota and Wagner 2006). In a sense, moths are actors and actresses that excel in mimicry, and their award is survival in the face of danger, despite their small size. Moths are an example of the outstanding properties and capabilities of nature, and this chapter will explore yet another one of moths' natural wonders: the surface of their eyes. Moth eyes' special nano-structured surface has garnered much attention from the scientific community, and you will be able to see just why once you read on.

Basic Structure of Moth Eyes

Optical System

Classified as having optical superposition eyes, moths use several facet lenses to allow light to reach the rhabdom, which is a long, cylindrical structure that has the visual pigment molecules of photoreceptors. This optical system makes moths much more sensitive to light compared to butterflies, who have apposition eyes that channel light into the rhabdom through a single facet lens and crystalline cone (Stavenga et al. 2005).

General Morphology

The eyes of the African Armyworm Moth (*Spodoptera exempta*) are hemispherical with a diameter of about 1 mm (Meinecke 1981), which gives an idea of general size and scale. Moths have compound eyes, which can be seen in Fig. 7.2, meaning they

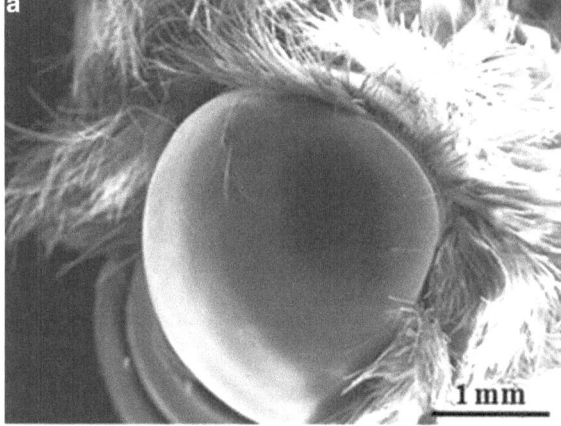
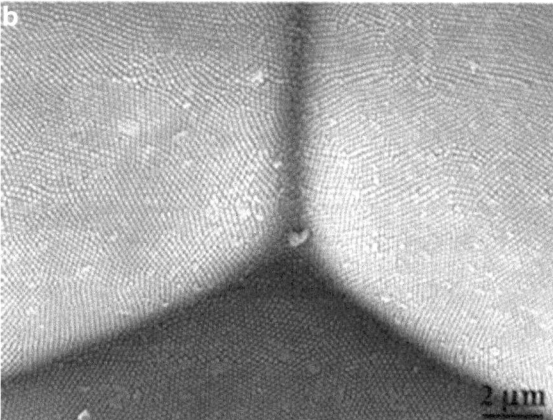

Fig. 7.2 A moth eye as seen in scanning electron microscopy (SEM) images with scale bars (**a**) 1 mm and (**b**) 2 μm (Reprinted from Phillips and Jiang (2013). With permission from Elsevier)

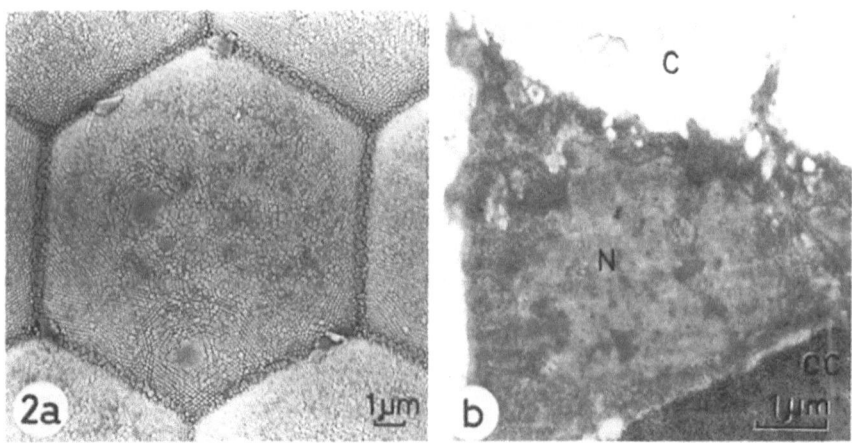

Fig. 7.3 (a) The array of corneal nipples can be seen on the surface of a moth's eye. (b) A crystalline cone (*CC*) cell, the cornea (*C*), and the nucleus (*N*)

are faceted and consist of many repeated, anatomically identical units called ommatidia (Stavenga et al. 2005). The number of ommatidia depends on the type of moth: *Spodoptera exempta* has about 8,000 ommatidia on its eye (Meinecke 1981), while *Antheraea polyphemus* has about 10,000 (Anton-Erxleben and Langer 1988). Every ommatidium detects signals that are neurologically processed to form a whole image. Each is composed of retinula cells, a rhabdom, a crystalline cone, and a corneal lens covered in nano-scale structures called corneal nipples (Lee and Erb 2013), which will be discussed in detail later in the chapter. The crystalline cone is composed of crystalline cone cells, whose cytoplasm is granular, electron dense, and lacks organelles except a few mitochondria and a nucleus (Meinecke 1981). See Fig. 7.3 for an SEM image of the array of corneal nipples and a crystalline cone cell.

Corneal Nipples

Antireflective Mechanism

Corneal nipples, which cover the corneal lens as mentioned earlier, were first observed and termed in 1962 by Bernhard and Miller, who discovered the regularly arrayed set of cuticular protuberances (Stavenga et al. 2005). Corneal nipples are often found in ordered hexagonal arrays (Meinecke 1981), where the spacing between them ranges from 180 to 240 nm, while their height varies from 0 to 230 nm (Yamada et al. 2010).

Slightly tapered from bottom to top, these protuberances are responsible for the antireflective property of moth eyes. The gradual change in diameter of the nanostructure provides a smooth transition for light as it transfers from air to the surface

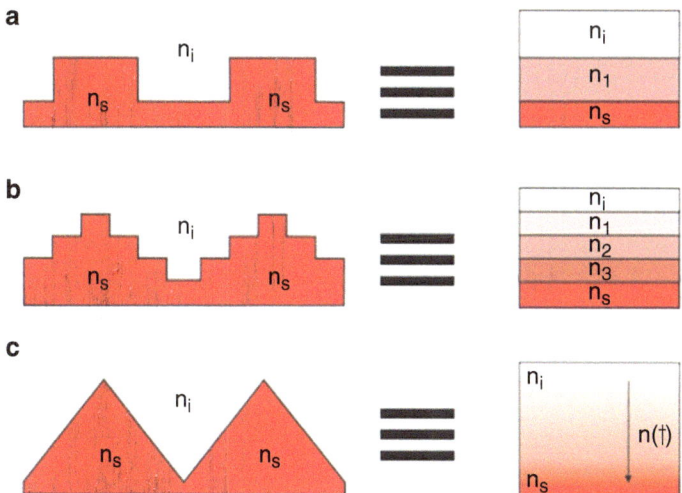

Fig. 7.4 Analogous profiles of refractive indices with their respective surface profiles consisting of (**a**) ridges, (**b**) steps, and (**c**) triangles (Reproduced with kind permission from Boden (2009))

of each ommatidium, which is made of chitin. Each incident photon encounters the slimmer tops of the protuberances first, causing the effective index of refraction to be only slightly higher than that of air. Then, as the protuberance increases in diameter and the photons get closer to the bottom, the index of refraction approaches that of chitin (Parker 1999). This optical mechanism relies on the fact that the wavelength range of incident light is less than the dimensions of the corneal nipples (Boden and Bagnall 2012). Therefore, despite the large difference between the indices of refraction of air and chitin, which are 1 and 1.55 respectively (Parker 1999), the smooth transition such as in Fig. 7.4c prevents light from encountering an abrupt change in refractive index such as in Fig. 7.4a, b, which would cause part of the light to be reflected. A helpful way to think of this transition is considering the tapered profile as infinitesimal layers that are infinitely stacked allowing the refractive index to change gradually and ever so slightly between each layer (Boden and Bagnall 2012).

Purposes of Antireflective Corneal Nipple Array

The antireflective surface of moth eyes is critical to their survival, particularly at night. By significantly reducing the amount of light reflected from moths' eyes, corneal nipples help moths be stealthier and less visible to potential predators in the night, since glints from their eyes will not be giving them away (The Optical Society 2012). Remarkably, this array of protuberances is not limited to moth eyes. They have been found on moth wings as well, allowing them to camouflage with their surroundings. *Cryptotympana aquila* and *Cephonodes hylas* are examples of moths who have this feature on their wings; the nipple arrays on the wing of *Cephonodes*

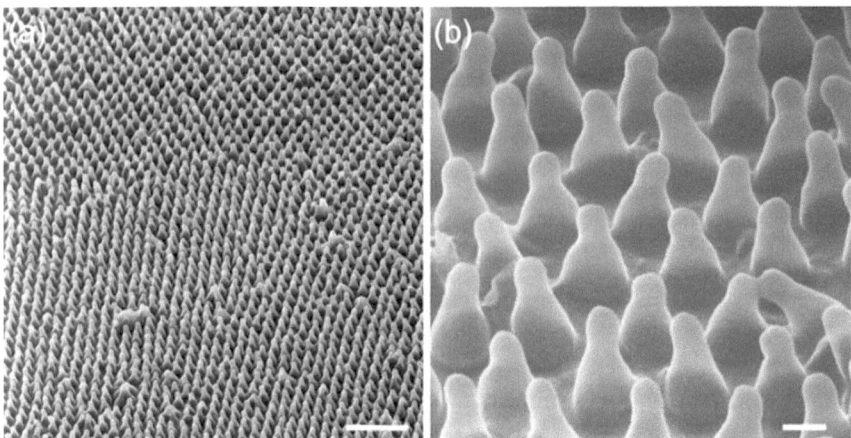

Fig. 7.5 Scanning helium ion microscope images showing (**a**) corneal nipple protuberances on the wing of *Cephonodes hylas* (scale bar: 1 μm) and (**b**) magnified version (scale bar: 100 nm) (Reprinted from Boden and Bagnall (2012), Fig. 1. With kind permission from Springer Science + Business Media)

hylas can be seen in Fig. 7.5. The array of corneal nipples can be found on both the dorsal and ventral sections of their wings. As a result, moths are unseen by prowling predators, with optical transmission over 90 % for a wide range of wavelengths (Boden and Bagnall 2012).

Corneal nipples give moths other advantages in addition to heightened safety from predators. The high transmission of light aids with low-light vision, which is particularly useful for nocturnal moths (Lee and Erb 2013). Overall, the reduction in reflectivity and increase in light absorption enhances the quality and sensitivity of the moth's visual system. Lastly, the nano-scale protuberances impart superhydrophobic properties, lending anti-adhesive, anti-fogging, and self-cleaning abilities (Boden and Bagnall 2012).

Engineering Applications

Moth-Eye Antireflective (AR) Structures

Researchers and scientists have utilized biomimicry of the moth-eye nano-structures to enhance the surfaces of materials such as glass and silicon to an extent that surpasses the antireflective properties provided by traditional antireflection coatings, or ARCs. The superior antireflective performance of moth-eye structured surfaces can be seen in Fig. 7.6, where a comparison of silicon with fabricated moth-eye structures, bare silicon, and thin-film ARCs shows that the application of moth-eye structures generated less reflectance over a wide range of wavelengths and angles of incidence (Boden 2009).

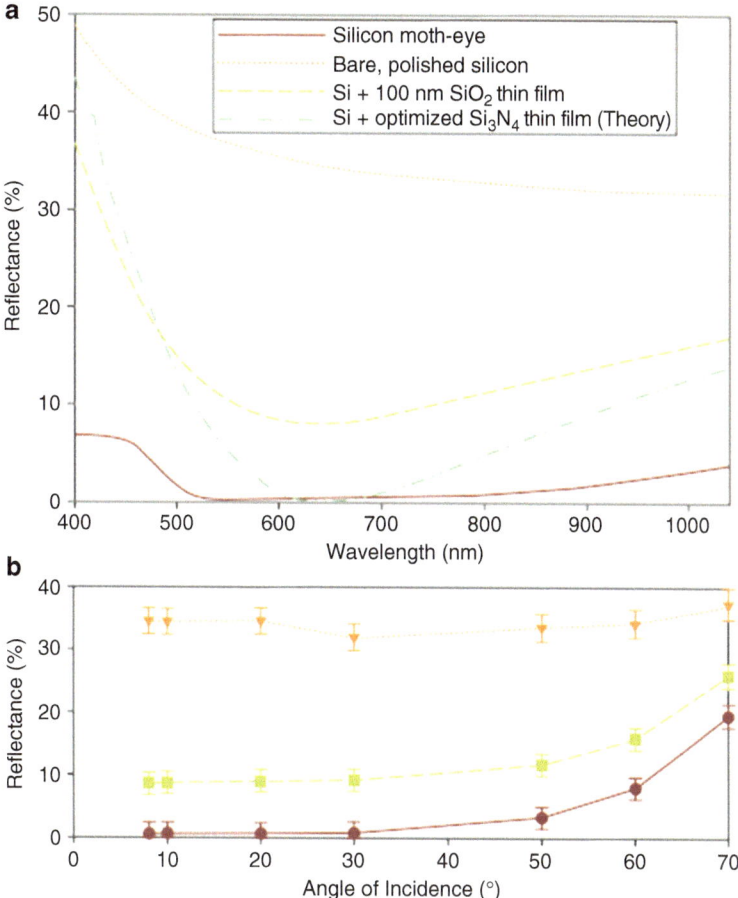

Fig. 7.6 A comparison of silicon with moth-eye surface structure to thin-film anti-reflection coatings (ARCs) and bare silicon: (**a**) reflectance versus wavelength and (**b**) reflectance versus angle of incidence (Reprinted from Boden and Bagnall (2012), Fig. 8. With kind permission from Springer Science + Business Media)

However, the corneal nipple pattern, which is on the scale of several hundred nanometers, is too small to be manufactured using standard photolithography techniques, requiring other processes such as electron-beam lithography and nanoimprint lithography (Boden and Bagnall 2012). Using a combination of injection-molding and hot-process embossing, the Fraunhofer Institute in Germany fabricated glass that has visual transmission of over 98 % and plastic of over 99 % (Leydecker 2008).

Interference lithography is another process that has been developed to obtain the moth-eye-like antireflective surface structure, like the one in Fig. 7.7. Clapham and Hutley were the first to demonstrate fabrication of artificial moth eyes by interference lithography, which can now be used to produce replications on a large,

Fig. 7.7 Fabricated moth eye replica as seen in scanning electron microscopy (SEM) micrograph (Reprinted from Forberich et al. (2008). With permission from Elsevier)

industrial scale. In addition, the moth-eye surface can be produced on a wide range of flexible substrates, which makes the surface structure viable for solar cells, among many other applications (Forberich et al. 2008).

Efficient Solar Energy Capture

Researchers in the solar energy field have taken particular interest in the antireflective properties of moth eyes. Photovoltaic (PV) systems, the processes of which can be seen in Fig. 7.8, are considered to be a powerful potential option in remediating the environmental and energy crisis that looms before us. However, because photovoltaic (PV) solar cells are made of materials that have high indices of refraction such as Si, GaAs, and InP, the absence of antireflective technology will allow them to lose more than 35 %. Ideally, a light-efficient solar energy cell does not waste energy, so minimizing optical reflection is an important challenge to overcome in order to obtain maximum benefits. Because a reduction in optical reflection can lead to a direct increase in the efficiency of such systems, applying anti-reflective techniques to PV cells has become of utmost importance (Yamada et al. 2010).

One concern that has challenged researchers is the robustness of potential moth-eye nano-structures. Given that solar cells will be continuously exposed to outdoor conditions, the question of whether fabricated structures will be able to endure the abrasion that harsh conditions may create remains. Therefore, the mechanical

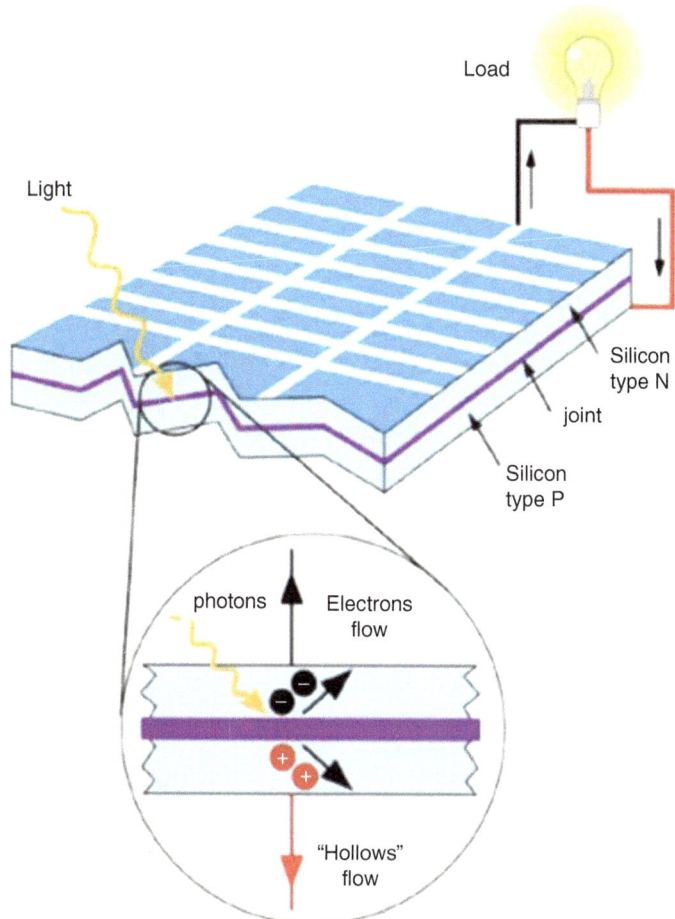

Fig. 7.8 The processes of a silicon photovoltaic cell (PV) cell (Reprinted from Papadopoulou (2011), Fig. 4.4. With kind permission from Springer Science + Business Media)

properties of the substrate are essential factors when it comes to producing moth-eye structures on materials that will be subject to such conditions (Boden and Bagnall 2012). If success in creating solar energy cells textured with the moth-eye surface could be achieved, however, the overall performance of PV systems can be improved by as much as 15 % (Leydecker 2008).

X-Rays of the Future: Safer and High-Performance

Perhaps one of the most exciting, up-and-coming applications of the moth-eye nano-structure is an X-ray enhanced with a film of protuberances modeled after corneal nipples. Yasha Yi, a professor at the City University of New York, and his colleagues are taking a look at the potential improvements that applying a thin, moth-eye-like film to X-rays can have on its performance. In short, X-rays work by absorbing the energy of X-ray photons and reemitting the energy in light form, which is then detected and formed into an image. The material that performs this conversion is called a scintillator. In order to improve the performance of X-rays—that is, increase the intensity of emitted light signals that is detected and formed into an image—there are two options: increase the dosage of X-ray light or increase the efficiency of X-ray to light conversion. The first option, however, is dangerous for patients, as they would be exposed to increased doses of harmful radiation. Therefore, Yi et al. aimed to utilize the second method (The Optical Society 2012).

Specifically, they added a thin film of cerium-doped lutetium oxyorthosilicate with pyramid-shaped protuberances resembling corneal nipples onto the scintillator of an X-ray. Even the density of the fabricated protuberances—100,000 to 200,000 in a 100×100 µm square—resembled that of a real moth eye. Experiments comparing a scintillator with the moth-eye film addition to a traditional scintillator showed a 175 % increase in the intensity of emitted light from the new set-up. According to Yi, much more work needs to be done, but potential achievements with the moth-eye based film could include lower doses of radiation and higher-resolution imaging—both things that would be a huge step forward for medical care and, best of all, patient treatment (The Optical Society 2012).

The surface structure and antireflective mechanism of moth eyes is a true natural wonder. Just as the antireflection and textured surface can save moths' lives by protecting them from predators at night, applying the technology to X-rays may save human patients and enhance their quality of life through higher-resolution imaging. Furthermore, significant improvements in PV solar cells can be made with moth eye surface replication. Though tiny, moth eyes have the potential to contribute enormously to engineering endeavors, serving as a reminder of the infinite amount of knowledge hidden even among nature's smallest creations.

References

Anton-Erxleben F, Langer H (1988) Functional morphology of the ommatidi in the compound eye of the moth, Antheraea polyphemus (Insecta, Saturniidae). Cell Tissue Res 252:385–396

Boden SA (2009) Biomimetic nanostructured surfaces for antireflection in photovoltaics. PhD thesis, University of Southampton, School of Electronics & Computer Science, p 18

Boden SA, Bagnall DM (2012) Moth-eye antireflective structures. In: Bhushan B (ed) Encyclopedia of nanotechnology. Springer, Netherlands, pp 1467–1477. doi:10.1007/978-90-481-9751-4_262

Forberich K, Dennler G, Scharber MC, Hingerl K, Fromherz T, Brabec CJ (2008) Performance improvement of organic solar cells with moth eye anti-reflection coating. Thin Solid Films 516(20):7167–7170. doi:10.1016/j.tsf.2007.12.088

Gray R (2010) Butterflies and moths mimic snakes and foxes to fool predators, claims researcher. http://www.telegraph.co.uk/earth/wildlife/8082739/Butterflies-and-moths-mimic-snakes-and-foxes-to-fool-predators-claims-researcher.html. Accessed 24 Aug 2013

Lee KC, Erb U (2013) Grain boundaries and coincidence site lattices in the corneal nanonipple structure of the Mourning Cloak butterfly. Beilstein J Nanotechnol 4:292–299. doi:10.3762/bjnano.4.32

Leydecker S (2008) Nano materials in architecture, interior architecture and design. Birkhäuser, Basel. doi:10.1007/978-3-7643-8321-3

Meinecke CC (1981) The fine structure of the compound eye of the African Armyworm Moth, Spodoptera exempta Walk. (Lepidoptera, Noctuidae). Cell Tissue Res 216:333–347. doi:10.1007/BF00233623

Papadopoulou EVM (2011) Photovoltaic industrial systems. Photovoltaic energy. Springer, Berlin/ Heidelberg, pp 31–55

Parker AR (1999) Light-reflection strategies: natural selection has produced a wealth of surfaces that interact efficiently with light. Technological applications abound, from better windows to stealth. Am Sci 87(3):248–255

Phillips BM, Jiang P (2013) Chapter 12: Biomimetic antireflection surfaces. In: Lakhtakia A, Martín-Palma RJ (eds) Engineered biomimicry. Elsevier, Boston, pp 305–331. doi:10.1016/ B978-0-12-415995-2.00012-X

Rota J, Wagner DL (2006) Predator mimicry: metalmark moths mimic their jumping spider predators. PLoS One 1(1):e45. doi:10.1371/journal.pone.0000045

Stavenga DG, Foletti S, Palasantzas G, Arikawa K (2005) Light on the moth-eye corneal nipple array of butterflies. Proc Biol Sci 273(1587):661–667. doi:10.1098/rspb.2005.3369

The Optical Society (2012) Insects inspire X-ray improvements: nanostructures modeled after moth eyes may enhance medical imaging. http://www.osa.org/en-us/about_osa/newsroom/ newsreleases/2012/insects_inspire_x-ray_improvements_nanostructures/. Accessed 29 July 2013

Yamada N, Kim ON, Tokimitsu T, Nakai Y, Masuda H (2010) Optimization of anti-reflection moth-eye structures for use in crystalline silicon solar cells. Prog Photovolt Res Appl 19(2):134–140. doi:10.1002/pip.994

Botanical Leaves: Groovy Terrain

Ignacio Estrada

Breathtaking Rice Terraces

The lush, vivid green rice terraces in Indonesia (Fig. 8.1) form breathtaking, awe-inspiring sights. Expansive fields overflow with rice leaves bathed in sunlight and dew, creating a peaceful landscape sure to be memorable for years and years. While the rice fields of Indonesia offer a relaxing, beautiful sight for travelers to gaze upon, they have produced the staple food for the native people for decades. Now, the rice leaves found in these terraces are known for even more: their unique surface properties related to superhydrophobicity and their bioinspirational potential.

Living nature serves as an inspiration for many innovations and continues to be an invaluable resource to solve technical challenges. Hydrophobic surfaces are no exception to such sources of inspiration, and as we can see from Fig. 8.2, the amount of research regarding hydrophobic surfaces has been increasing steadily since 2005 (Guo et al. 2011). Hydrophobicity has become more and more of an important topic over recent years given the variety of potential applications that such a technology would have—some of which have been explored earlier in this book in the chapter on lotus leaves. However, other botanical leaves such as rice leaves exhibit this special characteristic as well, and it is worthwhile to draw similarities and differences between the two types of leaves to gain even more insight into how superhydrophobicity works and its beneficial applications.

I. Estrada (✉)
e-mail: i.estrada.garcia@gmail.com

M. Lee (ed.), *Remarkable Natural Material Surfaces and Their Engineering Potential,*
DOI 10.1007/978-3-319-03125-5_8, © Springer International Publishing Switzerland 2014

Fig. 8.1 Rice terrace in Indonesia (Photo by Professor Q. Wang)

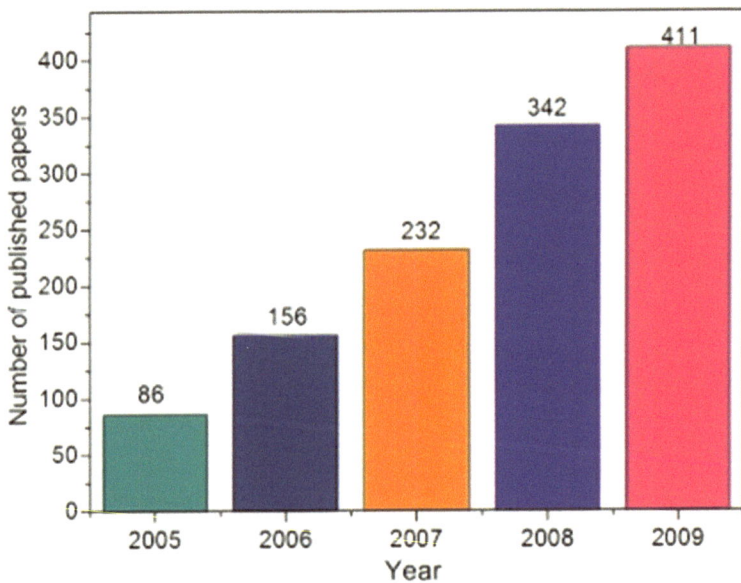

Fig. 8.2 The amount of published papers related to "superhydrophobicity" on the rise in ISI from 2005 to 2009 (Reprinted from Guo et al. (2011). With permission from Elsevier)

Rice Leaves

What Makes a Surface Hydrophobic?

Rice leaves repel water so efficiently that they are considered to be one of the few superhydrophobic materials (Feng et al. 2002). These properties are very useful in a variety of applications, which is why scientists have been trying to mimic the surface of rice leaves to create water repellent materials for a wide range of uses such as concrete and antenna coatings.

To define superhydrophobicity, we can take a look at two characteristics, the first being contact angle, which has been previously mentioned in the lotus leaf chapter. To briefly review the concept, when a liquid droplet comes in contact with a solid surface, it will either remain as a droplet or spread out on the surface to form a liquid film; such behavior is determined by what we define as the equilibrium contact angle. If the contact angle is greater than 90°, then the surface is considered to be hydrophobic. However, if the angle is above 150°, then the surface is considered to have one of the characteristics necessary for superhydrophobicity (Wang and Jiang 2007). If the contact angle of a water droplet on a surface is almost 0°, it is considered to be a superhydrophilic surface.

Another crucial factor that makes rice leaves so efficient at remaining dry is a low sliding angle, α, where α can be expressed as the difference between the advancing contact angle and the receding contact angle. This parameter, also known as contact angle hysteresis, is crucial because it determines the mobility of the water droplet and whether it will "stick" or roll down the surface. According to Johnson et al., the lower the value of the sliding angle, the easier it becomes for the water droplet to move on the surface (Johnson and Dettre 1964).

Therefore, rice leaf surfaces, which are superhydrophobic, are characterized by a contact angle greater than 150° and a sliding angle of less than 10° (Wang and Jiang 2007). The surface morphology of rice leaves is such that it exhibits both of these superhydrophobic characteristics. However, this behavior is only observed on the top surface of the leaves and not the bottom surface.

Surface Topology

To understand how rice leaves achieve their superhydrophobic properties, we must examine what the surface's physical and chemical characteristics are and how they contribute to the hydrophobic effect. If we were to examine a rice leaf with our bare eyes, we would just see a green surface that resembles any other plant leaf. Therefore, it must be the microstructure of the surface that determines the superhydrophobic properties of the leaf.

The first important characteristic of rice leaves is the distinct difference between the top and bottom surfaces of the leaves. While the top of the leaf is very smooth

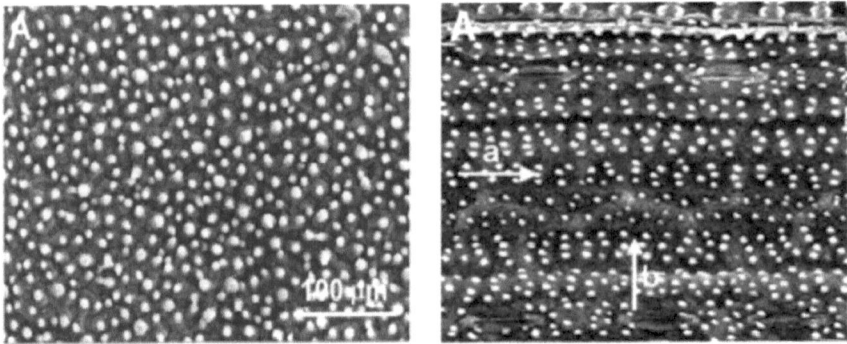

Fig. 8.3 Microstructure arrangement of micropapilla on rice leaf (*right*), and microstructure arrangement of micropapilla on lotus leaf (*left*) (Reprinted from Sun et al. (2005). With permission from American Chemical Society)

and soft to the touch, the bottom surface is a lot rougher. If we then looked at the two surfaces of the rice leaf under a microscope, we would start to see the major differences in their nano- and micro-structures that are responsible for the surfaces' different responses to water.

Furthermore, it becomes of vital importance to compare the rice leaf's properties with other similar surfaces such as lotus leaves in order to understand how different surface topologies affect the properties of the surface. Even though both leaves exhibit superhydrophobic properties there is one big difference: the lotus leaves show isotropic properties while the rice leaf exhibits anisotropic properties. If we look at both surfaces at the micro-scale level, we can appreciate a number of similarities and differences.

As evident in Fig. 8.3 both surfaces have a series of micropapillae on the surface of the leaf. These papillae secrete epicuticular wax crystals that, according to Barthlott et al., cause effective water repellency, minimizing the contact of water droplets with the surface of the leaf and facilitating the droplets to roll off. However, the arrangement of the papillae on the surface of the lotus leaf is random, while on the rice leaf, these papillae are arranged in such a manner as to create channels. These channels lead to a preferred rolling direction of the water droplets, resulting in the anisotropic properties of the rice leaf (Barthlott and Neinhuis 1997).

However, numerical calculations performed on these surfaces show that the contact angle could only reach a theoretical maximum of 147° (Feng et al. 2002), which is not enough to describe the superhydrophobic properties of these surfaces. Therefore, we must delve ourselves even deeper into the surface structure to understand how superhydrophobicity is achieved. Recent studies have shown nano-scale hierarchical structures on the leaf. The papillae were found to consist of further branch-like nano-structures that could contribute to higher contact angles (Feng et al. 2002). The nanostructure of the papillae can be seen in Fig. 8.4. This time, the comparison between lotus and rice leaves does not need to be made given that both papillae structures are found to be almost identical.

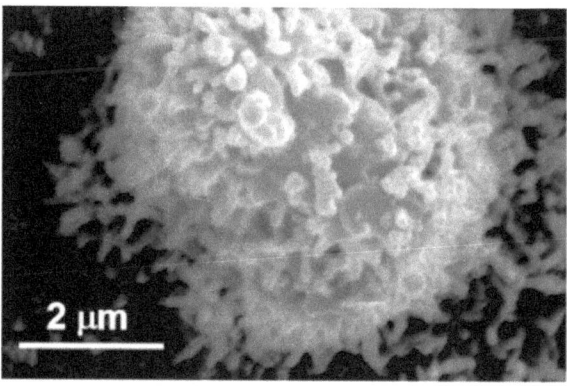

Fig. 8.4 Nanostructure of a rice leaf papillae (Reprinted from Sun et al. (2005). With permission from American Chemical Society)

Taking into account the effects created by the nanostructures on the papillae, it has been found that the contact angle may increase to about 160°, which is consistent with the experimental values found by Woodward et al. of $161.0° \pm 2.7°$ (Woodward et al. 2000). To further understand why the papillae provide such a high contact angle, we can examine Cassie's contact angle theory for rough surfaces based on Young's model for flat surfaces, as introduced in the lotus leaf chapter. According to this theory, the solid's surface is regarded as a solid-vapor composite interface instead of a flat surface, and the vapor pockets are assumed to be trapped underneath the liquid. This solid-vapor interface results in an enhancement in the contact angle between the water droplet and the surface if the angle is greater than 90°.

We can then conclude that it is indeed the micro- and nano-hierarchical structures of the surface combined with the surface chemistry of the wax that provides the superhydrophobic effect on rice leaves in addition to lotus leaves. However, the particular arrangement of the papillae on the rice leaf provides it with its unique anisotropic properties (Qu et al. 2010).

What makes rice leaves unique with respect to many other superhydrophobic surfaces, as we discussed before, is the anisotropic nature of it. When we examine the rice leaves, we can see how the roll-off angle—the critical angle at which the droplet begins to roll—is significantly lower in the longitudinal direction of the leaf than in the perpendicular direction of the leaf. This anisotropy originates from the multi-scale roughness of the surface (Lee et al. 2012). Figure 8.5 shows that the rice leaf has many parallel ridges or grooves and wavy layers, and further magnification of the leaf shows that papillae are arranged in a direction parallel to the ridge direction. Due to this structure, the water drop can roll freely in the longitudinal direction but is discouraged from running along the perpendicular direction. This effect can be characterized by the sliding angles in both directions, which is much lower for the longitudinal direction (<3–5°) than for the perpendicular direction (<9–15°) (Lee et al. 2012).

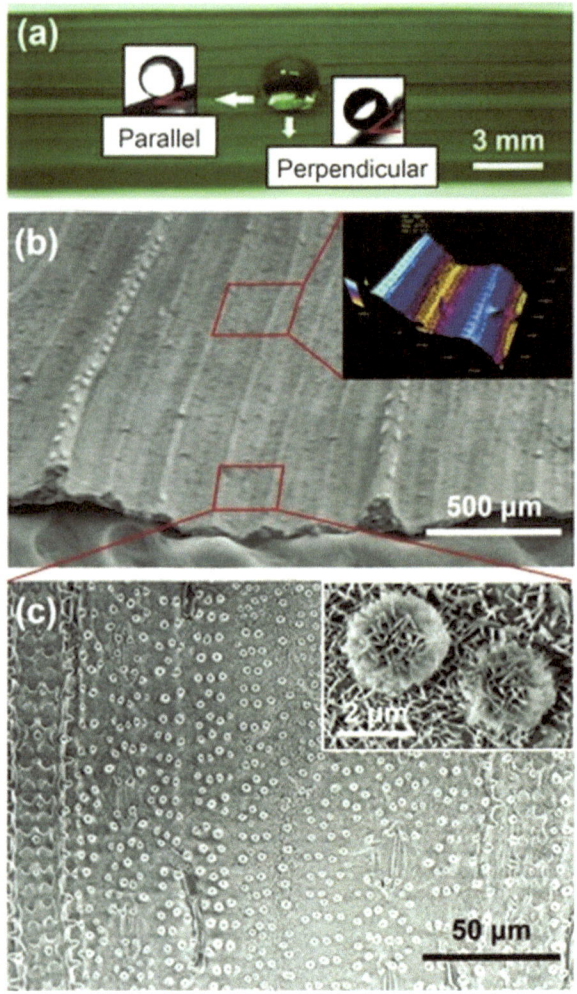

Fig. 8.5 (**a**) A rice leaf with a drop of water; insets depict roll-off angles parallel or perpendicular to the longitudinal direction of the leaf. (**b**) Scanning electron micrograph (SEM) of a rice leaf's upper surface; *inset* shows confocal laser scanning microscopy (CLSM) image of surface. (**c**) Image (b) magnified; *inset* shows papillae and a covering of wax (Reprinted from Lee et al. (2012). With permission from John Wiley and Sons)

Engineering Applications

Self-Cleaning Glass

Scientists have been able to replicate the superhydrophobic and anisotropic properties exhibited by rice leaves and apply them to different materials in order to achieve new materials that behave in ways we had never dreamed of before. One of the most widespread uses of superhydrophobic materials is the use of self-cleaning glass. In this application, glass is treated and coated using physical techniques such as ion etching and chemical techniques such as plasma-chemical roughening to replicate the pattern of rice and lotus leaves (Park et al. 2012). Once the glass is coated, it exhibits superhydrophobic properties and can be used in a variety of applications, such as the windows of skyscrapers where the outer surface of the glass is often hard to access. Instead of having to manually clean the windows, rainwater droplets can roll down the glass, dragging any type of dirt or dust with them and making these windows almost maintenance free.

Another use of self-cleaning glass is in the encapsulation of solar cells. A frequent problem with solar cells is that they get covered in dust or dirt, losing efficiency over time. By using self-cleaning glass, the need to clean them regularly is eliminated, reducing maintenance costs and increasing the life of the solar cells.

Additional Current Applications

Engineering uses for superhydrophobic coatings can also be found in the field of civil engineering and, more specifically, concrete. The main mechanism by which concrete structures get damaged over time is the ingress of water into the structure. This leads to corrosion of the concrete and even the steel reinforcement embedded in the concrete structure. For this reason, most outdoor concrete structures are coated with hydrophobic materials to effectively prevent such damages. These kinds of treatments are widely used because of the relative ease of the process and the fact that the appearance of the structure remains unchanged (Raupach and Büttner 2009).

Another well-known application of superhydrophobic coatings is in microwave and radio antennas. A common problem with antennas is the effect of rain on their performance. Water has a very high dielectric constant, and even thin films of water on the surface of antennas can cause a large attenuation of transmitted or received signals. Therefore, the use of superhydrophobic coatings on antennas is a widespread practice to improve the efficiency and reliability of telecommunication systems, as evident in Fig. 8.6 (Antonini et al. 2011).

Future Applications

Even though scientists are already using superhydrophobic surfaces and coatings in numerous technological applications, there are multiple avenues that have not been fully explored yet. Superhydrophobic materials could be used in ways we cannot even imagine today, and there are multiple experimental technologies that have already shown very promising results.

An example of a new use for superhydrophobic materials is being researched currently at the Massachusetts Institute of Technology, or MIT. Currently, a team of nanomaterial scientists and engineers at MIT has created a superhydrophobic material that is 10,000 times more slippery than any other existing surface today.

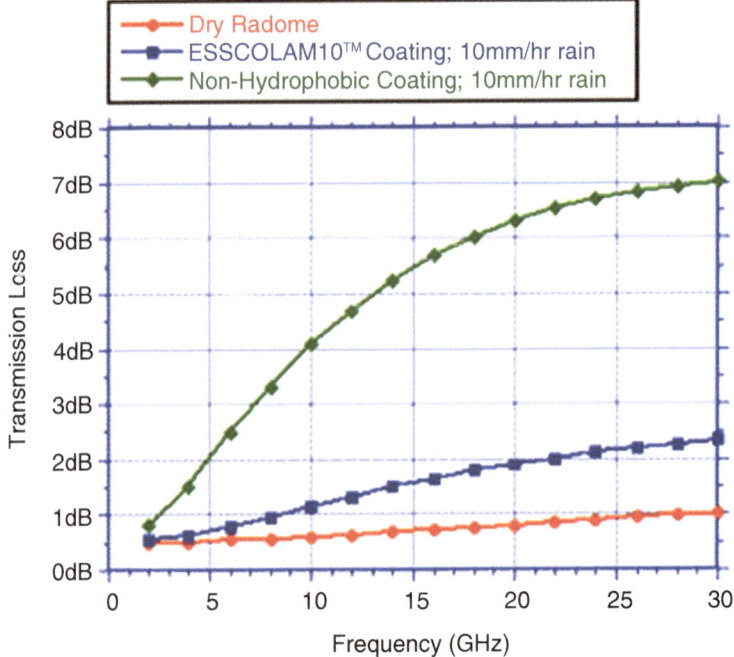

Fig. 8.6 Transmission loss for a microwave antenna: (**a**) that is dry (*red*, *dot*), (**b**) coated in super-hydrophobic coating (*blue*, *square*) (transmission loss when wet with coating), and (**c**) without superhydrophobic coating (*green*, *diamond*) (transmission loss when wet without coating) (Reprinted with kind permission from L-3 ESSCO (2013))

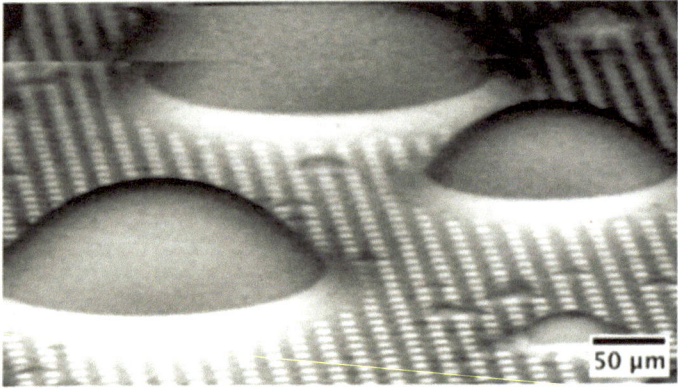

Fig. 8.7 MIT's oil lubricated nano structure (Reprinted with permission from Enhanced Condensation on Lubricant-Impregnated Nanotextured Surfaces © 2012 American Chemical Society)

Scientists have been able to create a material with a surface topology such that it can hold by means of capillary forces an oil lubricant particle in the gaps between the protuberances in the surface texture. Once the material, pictured in Fig. 8.7, has been dipped in lubricant, the lubricant remains fixed to the material, greatly

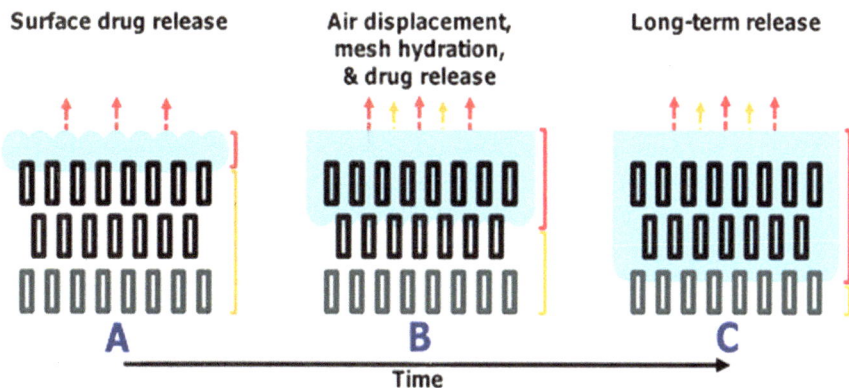

Fig. 8.8 Potential novel drug-releasing superhydrophobic material (Reprinted from Yohe et al. (2012). With permission from American Chemical Society)

increasing water droplet mobility on the surface. This material could have count-less applications, but a very important one would be to coat the inside of cooling towers in power plants, reducing the moisture build up and increasing the effi-ciency of the turbine. This would have a dramatic effect on society, since almost 80 % of the world's power production utilizes steam turbines and therefore requires cooling towers (Anand et al. 2012).

Another potential use of superhydrophobic materials comes from the world of medicine, where a new application involving the way drugs are delivered inside the human body is being developed. Right now there are several unmet clinical needs for a way to deliver a drug over extended periods of time (weeks). Professor Mark Grinstaff at Boston University has developed an experimental superhydrophobic coat-ing to control the rate of drug release. This is possible because superhydrophobic materials trap air between the protuberances in their surfaces. This trapped air in turn prevents liquids from penetrating the surface, and only after the drug is immersed in a liquid such as blood will the air gradually be displaced, releasing the drug (refer to Fig. 8.8). Experimental data shows that the release time of an anti-cancer pill coated with the new hydrophobic material increased fourfold—from 15 days for a pill with-out any coating to 70 days. If this coating proves to be effective on other drugs, it may very well revolutionize how drugs are administered and, ultimately, patient well-being (Yohe et al. 2012). Furthermore, superhydrophobic materials can have an impact in the field of micro-machines. Recent research has proven possible the creation of a simple and compact fuel-cell consisting only of solid channels. In this case, a super-hydrophobic film is used to facilitate the removal of byproduct CO_2 without the need of any discrete pump or gas separator. The superhydrophobic membrane allows the CO_2 to escape through the numerous pores in the venting membrane without disrupt-ing the flow of the fuel. This fuel cell design avoids the problem known as packaging penalty, which has been hindering the miniaturization of fuel cells below a centimeter. The creation of such a fuel cell is crucial in the process of developing micro-machines such as robots that could access otherwise inaccessible structures to diagnose or fix any damages (Hur et al. 2012).

Finally, superhydrophobic materials can be applied in long pipes and tubes. The transport of water and other Newtonian liquids through a smooth pipe is dominated by the frictional drag of the liquid against the walls. The resistance of flow can be reduced by creating a layer of gas between the fluid and the solid pipe. Current methods to do so involve a continuous energy input such as an air pump. However, research shows that the same effect can be achieved using superhydrophobic materials (Shirtcliffe et al. 2009). Furthermore, a more detailed control of the flow can be achieved through the use of anisotropic superhydrophobic materials, which would, in a way, guide the flow throughout the pipe or tube. Experimental results show how the use of a superhydrophobic coating could cause up to 20–30 % reduction in friction, which would have great benefits and improve the efficiency of countless systems (Cao 2005).

Through the study of the surface of rice and lotus leaves, we have gained a deeper understanding of the conditions necessary for superhydrophobic surfaces and the properties that such surfaces entail. We have seen the importance of these surfaces in modern day science and technology, as well as how they can help improve current systems or solve technological problems to impact society in a beneficial way.

References

Anand S, Paxson AT, Dhiman R, Smith JD, Varanasi KK (2012) Enhanced condensation on lubricant-impregnated nanotextured surfaces. ACS Nano 6(11):10122–10129. doi:10.1021/nn303867y

Antonini C, Innocenti M, Horn T, Marengo M, Amirfazli A (2011) Understanding the effect of superhydrophobic coatings on energy reduction in anti-icing systems. Cold Reg Sci Technol 67(1–2):58–67. doi:10.1016/j.coldregions.2011.02.006

Barthlott W, Neinhuis C (1997) Purity of the sacred lotus, or escape from contamination in biological surfaces. Planta 202(1):1–8. doi:10.1007/s004250050096

Cao LL (2005) Superhydrophobic surface: design, fabrication and applications. University of Pittsburg, Dissertation

Feng L, Li S, Li Y, Li H, Zhang L, Zhai J, Song Y, Liu B, Jiang L, Zhu D (2002) Superhydrophobic surfaces: from natural to artificial. Adv Mater 14(24):1857–1860. doi:10.1002/adma.200290020

Guo Z, Liu W, Su BL (2011) Superhydrophobic surfaces: from natural to biomimetic to functional. J Colloid Interface Sci 353(2):335–355. doi:10.1016/j.jcis.2010.08.047

Hur JI, Meng DD, Kim CJ (2012) Self-pumping membraneless miniature fuel cell with an air-breathing cathode. J Microelectromech S 21(2):476–483. doi:10.1109/JMEMS.2011.2176920

Johnson RE Jr, Dettre RH (1964) Contact angle hysteresis. III. Study of an idealized heterogeneous surface. J Phys Chem 68(7):1744–1750. doi:10.1021/j100789a012

L-3 ESSCO (2013) Resources: the importance of hydrophobic coating. http://www2.l-3com.com/essco/radomes/pages/the%20importance%20of%20hydrophobic%20coating.html. Accessed 22 Aug 2013

Lee SG, Lim HS, Lee DY, Kwak DH, Cho KW (2012) Tunable anisotropic wettability of rice leaf-like wavy surfaces. Adv Funct Mater 23(5):547–553. doi:10.1002/adfm.201201541

Park KC, Choi HJ, Chang CH, Cohen RE, McKinley GH, Barbastathis G (2012) Nanotextured silica surfaces with robust superhydrophobicity and omnidirectional broadband supertransmissivity. ACS Nano 6(5):3789–3799. doi:10.1021/nn301112t

Qu M, He J, Zhang J (2010) Superhydrophobicity, learn from the lotus leaf. In: Mukherjee A (ed) Biomimetics learning from nature. InTech, Shanghai. doi:10.5772/8789

Raupach M, Büttner T (2009) Hydrophobic treatments on concrete—evaluation of the durability and non-destructive testing. In: Alexander MG, Beushausen HD, Dehn F, Moyo P (eds) Concrete repair, rehabilitation and retrofitting II. Taylor & Francis Group, London, pp 907–913

Shirtcliffe NJ, McHale G, Newton MI, Zhang Y (2009) Superhydrophobic copper tubes with possible flow enhancement and drag reduction. ACS Appl Mater Interfaces 1(6):1316–1323. doi:10.1021/am9001937

Sun T, Feng L, Gao X, Jiang L (2005) Bioinspired surfaces with special wettability. Acc Chem Res 38(8):644–652. doi:10.1021/ar040224c

Wang S, Jiang L (2007) Definition of superhydrophobic states. Adv Mater 19(21):3423–3424. doi:10.1002/adma.200700934

Woodward JT, Gwin H, Schwartz DK (2000) Contact angles on surfaces with mesoscopic chemical heterogeneity. Langmuir 16(6):2957–2961. doi:10.1021/la991068z

Yohe ST, Colson YL, Grinstaff MW (2012) Superhydrophobic materials for tunable drug release: using displacement of air to control delivery rates. J Am Chem Soc 134(4):2016–2019. doi:10.1021/ja211148a

Snake Skin: Small Scales with a Large Scale Impact

9

Michelle Lee

Slithering with Precision and Agility

Did you know that of all animals, Americans find snakes to be the number one most terrifying creature? According to a recent poll, 21 % of respondents claim to be the most afraid of snakes (Public Policy Polling 2013). While the sight of this scaly creature and its signature smooth glide would most likely cause an onlooker to cringe, it—believe it or not—brings about great joy in others. Scientists have been intrigued for years by the snake's ability to slither along a remarkable variety of surfaces with precision and agility. Take, for example, Dr. Howie Choset of Carnegie Mellon University and the researchers of his Modular Snake Robot Lab. These 'Modsnakers', as they like to call themselves, create robotic snakes (pictured in Fig. 9.1) by modeling the snake's movements with numerous hinged joints and 16 degrees of freedom.

By imitating snake movement, the robots are able to travel along non-solid rubble surfaces as well as through tight or windy spaces that conventional robots may not be able to access. According to Carnegie Mellon News, this kind of agility in the snake robots make them ideal for aiding in search and rescue missions. They are particularly useful in scenarios involving radioactive environments, such as power plant facilities. The snake robots have already been sent through the pipes of a nuclear power plant to places that would be difficult, impossible, or hazardous to reach for humans (Spice 2013). As snake movement becomes a subject of increasingly expansive study, it is important to acknowledge one of the prime enablers of such movement: the snakes' scales themselves.

Particular gratitude to Joseph Park for allowing me to consult his draft on snake skin.

M. Lee (✉)
Mechanical Engineering, McCormick School of Engineering
at Northwestern University, Evanston, IL 60208, USA
e-mail: MichelleLee2013@u.northwestern.edu

M. Lee (ed.), *Remarkable Natural Material Surfaces and Their Engineering Potential,*   103
DOI 10.1007/978-3-319-03125-5_9, © Springer International Publishing Switzerland 2014

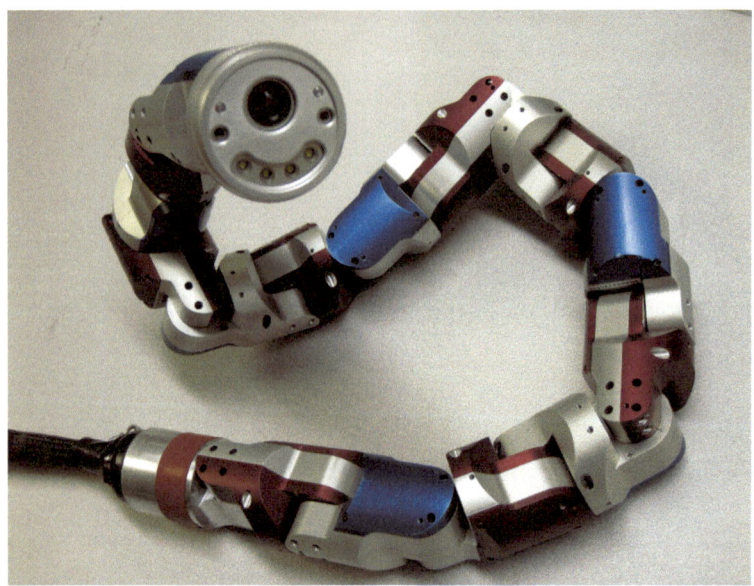

Fig. 9.1 Robotic snake created at the Modular Snake Robot Lab at Carnegie Mellon University (Reprinted with kind permission from Howie Choset (2010))

Frictional Anisostropy

What Is Frictional Anisotropy?

The forward movement of a snake is complex, involving interactions between numerous factors including, but not limited to, weight distribution, its mode of motion, the scales, and the characteristics of the surface it is moving on. As legless creatures, they rely on propelling themselves forwards by transmitting frictional forces to the ground through perfectly coordinated movements (Abdel-Aal and El Mansori 2009). Despite subjecting their scales to constant friction and contact with surfaces, snakes' skin remains remarkably intact. This resistance to wear and tear is a subject of many researchers' interest. This chapter's focus will be on the tribological properties of snake scales, in particular the ventral scales on the underside of the snake, as they are the most critical to movement and display an important characteristic that scientists call frictional anisotropy.

Frictional anisotropy can be thought of as directional friction, or varying coefficients of friction based on orientation. In an article published by Georgia Tech, it was described as the "resistance to sliding in certain directions" (Vogel 2009). Frictional anisotropy can be found in a plethora of natural and artificial systems, ranging from burr-covered plant leaves and global plate tectonics to the textures and structures of certain solids at the molecular level. In snakes, frictional anisotropy is considered to be a result of preferred orientation of the micro-scale surface

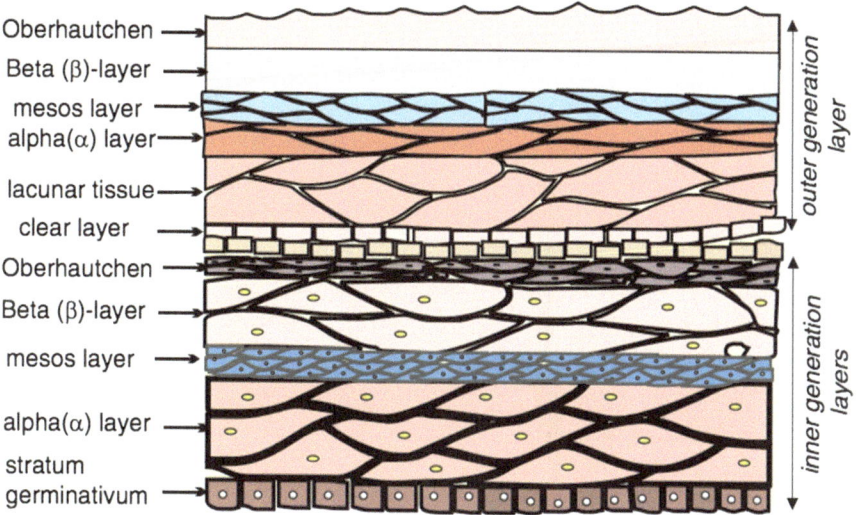

Fig. 9.2 Generalized epidermis of snake skin divided into the layer about to be shed (outer generation layer) and the layer to replace shed layer (inner generation layer) (Reprinted from Abdel-Aal and El Mansori (2011), Fig. 4.2. With kind permission from Springer Science + Business Media)

structures in addition to the scales themselves (Filippov and Gorb 2013). Frictional anisotropy plays a critical role in snake motion and the mechanics of its propulsion.

Two Reasons for Frictional Anisotropy of Snake Skin

The two general strata that make up snake skin are called the dermis and the epidermis. While the dermis consists of connective tissue and lies under the epidermis, the epidermis consists of seven layers and acts as an outer shield for the snake's body. As shown in Fig. 9.2, the top layer is called the Oberhautchen; it is composed of keratinized dead skin cells and is the toughest layer. Under the current epidermis, a snake develops a new epidermal layer that is attached to the original by a spinulae structure. Then, a widely known process called 'molting' occurs, where the snake sheds four layers of the older, outermost epidermal layer, which consist of the proteinaceous inner α–layer, the mesos layer rich in lipids, the proteinaceous β–layer, and the Oberhautchen layer (Abdel-Aal 2013).

As the outermost layer, the Oberhautchen is in direct contact with the environment, thus receiving direct wear and tear from the surfaces treaded by the snake. This layer contains hexagonal ventral scales of varying sizes connected by flexible skin (Abdel-Aal and El Mansori 2011). The hexagonal pattern of the snake's ventral scales, however, is not enough to explain the frictional anisotropy of the snake's

Fig. 9.3 Snake scales (Reprinted with kind permission from Lucy Browne (2012))

surface; instead, this directional friction can in part be attributed to the wide overlapping of the ventral scales (Fig. 9.3).

According to a study completed by Hu et al., these overlaps snag on asperities—
or roughness and unevenness—on the surface in the direction of the scale, creating
varying coefficients of friction in certain directions. It was found that the static coefficient of friction, μ, when the snake moved in the forward direction was generally
lower than that of the snake when it moved backwards. Based on the resulting values of μ, which was the highest moving flank-wise at $\mu_t = 0.20 \pm 0.015$, intermediate
towards the tail at $\mu_b = 0.14 \pm 0.015$, and lowest sliding forward at $\mu_f = 0.11 \pm 0.011$,
Hu et al. concluded that the snake had preferred directions for sliding given certain
surface characteristics (e.g. roughness level). The role of asperities in the frictional
anisotropy that snakes experience while sliding was confirmed when the researchers
evaluated coefficients for an entirely smooth surface. With almost no asperities
available for snagging, the coefficient of friction moving flank-wise was about
$\mu_t \approx 0.16$, while the coefficients of friction moving forwards and backwards were
about $\mu_f \approx \mu_b \approx 0.14$, showing no significant directional preference (Hu et al. 2009).

In addition to the overlapping of scales, micro-hairs known as microfibrils found
on the snake's scales are also responsible for frictional anisotropy. The microfibrils,
which measure about 30–100 nm in height and 100–400 nm in diameter, are regularly arranged, creating either row-like patterns or triangular arrays, and they are all
oriented in the sliding direction with tips elevated and pointing toward the tail. In
fact, the upward slope of the microfibrils causes their tips to be almost three to four
times more elevated than the cross-sectional heights (Hazel et al. 1999). As shown
in the Fig. 9.4 in which snake skin is subject to ×5,000 magnification, boundaries
between scales consist of flexible, soft tissue (left), and while the scales show
microfibrillar texture (right). Fibrillar arrays differ in shape and spacing based on
the region of the snake. Note the uniform direction of the microfibrils (Abdel-Aal
and El Mansori 2011).

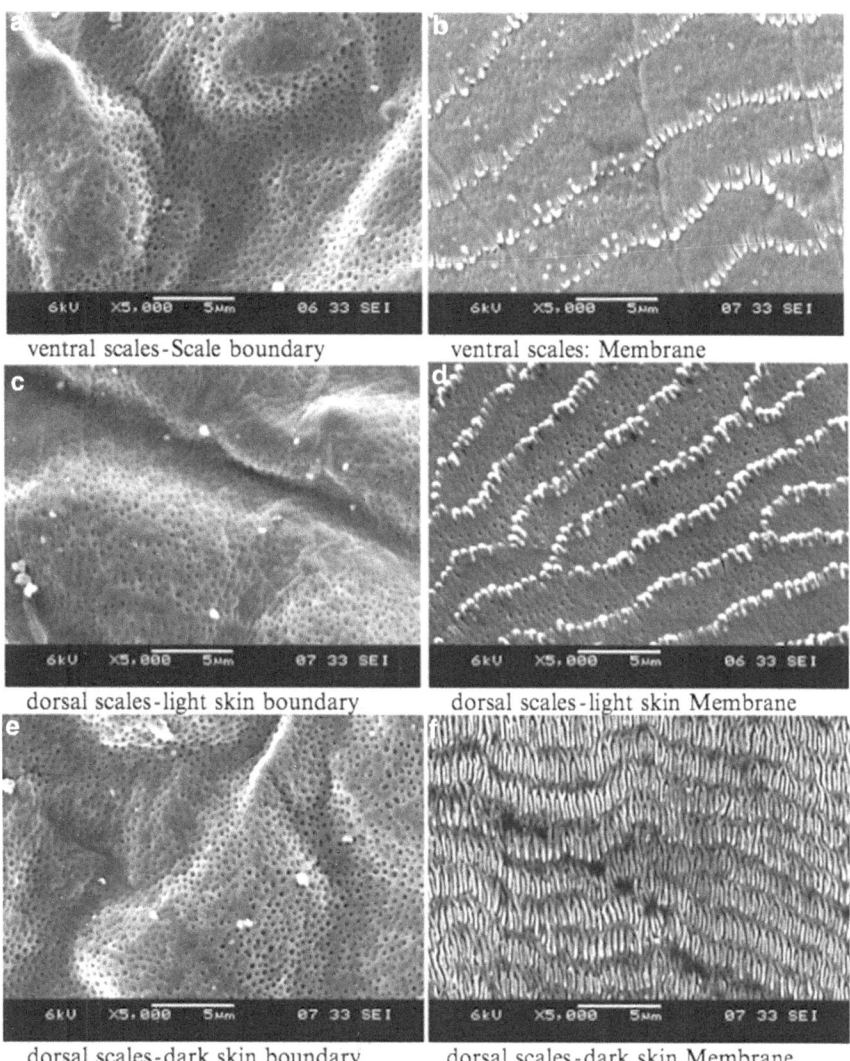

ventral scales-Scale boundary   ventral scales: Membrane

dorsal scales-light skin boundary   dorsal scales-light skin Membrane

dorsal scales-dark skin boundary   dorsal scales-dark skin Membrane

Fig. 9.4 Scanning electron microscopy (SEM) images reveal marked differences in boundary and membrane features: (**a, b**) show ventral scale boundaries and membrane, (**c, d**) show dorsal scale light boundaries and membrane, and (**e, f**) show dorsal scale dark boundaries and membrane. Scale: 5 μm; magnification: ×5,000 (Reprinted from Abdel-Aal and El Mansori (2011), Fig. 4.9. With kind permission from Springer Science + Business Media)

The orientation of these microfibrils and their significant upward slope incur friction force spikes that indicate frictional anisotropy. When atomic force microscopy (AFM) tips were dragged across the microfibrils of snake scales, Hazel et al. found that the friction force values differed greatly depending on the direction. Dragging the AFM tip in the reverse direction—indicated by the white arrow in

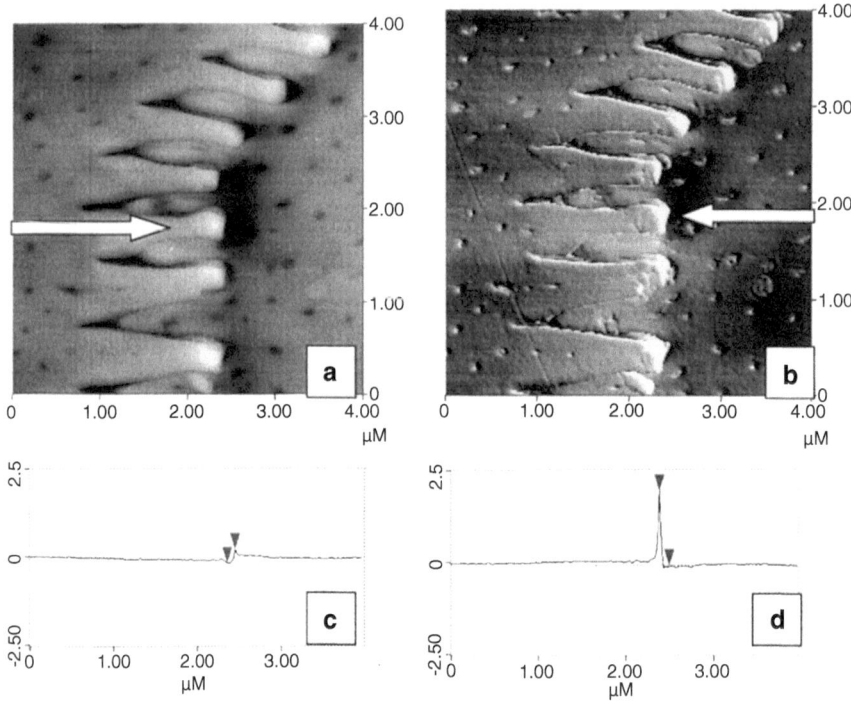

Fig. 9.5 4×4 μm friction images with arbitrary Z-range in which brighter denotes higher friction signal: (**a, c**) shows tip moving forward from *left to right* and the corresponding friction signal, while (**b, d**) shows tip moving backward from *right to left* and the corresponding friction signal (Reprinted from Hazel et al. (1999). With permission from Elsevier)

Fig. 9.5b—incurred a friction force spike four to six times greater than when the tip was dragged in the same direction as the microfibrils—indicated by the white arrow in Fig. 9.5a. Figure 9.5c, d display the contrast between the friction force spikes resulting from the two cases, confirming the microfibrillar texture and orientation as causes for the frictional anisotropy of snake scales (Hazel et al. 1999).

Remarkable Wear Resistance

Varying Load Bearing Capacities

Another interesting attribute of snake skin is its ability to resist wear despite being constantly subjected to contact with surfaces, even in particularly tribologically hostile environments (Abdel-Aal and El Mansori 2011). Like frictional anisotropy, the snake's exceptional resistance to wear is also due to both macro- and micro-level causes.

On the macro-level, observations of the overall scheme of ventral scale placement through the length of the snake have revealed non-uniformity. Because most of the snake's mass is concentrated in the trunk of the body and the head and tail are relatively skinny and light, in their research of the Ball python, Abdel-Aal and El Mansori found that the skin of the snake has adapted accordingly. Because the trunk is the heaviest and most used part of the body in terms of traction and rubbing during motion, it is expected to reflect the bulk of the reaction forces that the snake experiences in sliding.

Quantitative measurements of load bearing capacity were obtained by producing an Abbott-Firestone Load Curve (AFLC) of scale samples taken from three regions. The purpose of the AFLC is to obtain a particular surface's behavior and potential for damage due to wear when sliding. This is done by integrating the probability density distribution of the surface profile with respect to surface height.

Curve 1 in Fig. 9.6 corresponds to ventral scales taken from the trunk portion, curve 2 to those taken from the neck-trunk area, and curve 3 to those taken from the tail end. The plot of the resulting AFLC in Fig. 9.6 reveals a similar load bearing capacity for ventral scale samples from the throat-neck and tail sections, while the capacity for scales taken from the trunk is exceptionally high. It is no wonder, then, that the snake is able to withstand much exposure to rubbing and traction without significant wear. It seems as though it was designed by nature to have higher load bearing capacity in precisely the area that would receive the most loading. Thus, the probability of sustaining severe topographical damage from this type of loading is unlikely, as the areas with the highest load are associated with the scales that are naturally most able to handle it (Abdel-Aal and El Mansori 2011).

Double-Ridge Design of Microfibrils

In addition to placement along the body, snake scales are also microscopically adapted to resist wear. The previously mentioned microfibrils feature a double-ridge design that not only provides ideal conditions for sliding in the forward direction with minimum friction and adhesive force, but it also helps to effectively stop the snake's motion in the opposite direction. The double ridges run along the edges of the microfibrils and are of nano-scale geometry, measuring 3–5 nm in height. Using an atomic force microscopy (AFM) tip with a radius of 20–40 nm, Hazel et al. found the radius of curvature of the ridged ends of the microfibrils to be around 30–50 nm.

Though tiny in geometry, the ridges enhance snakes' sliding abilities by reducing friction and adhesive forces through minimizing contact area. According to Amontons' law of friction, the real contact area between two bodies is proportional to the amount of friction generated at the interface. It then follows that decreasing real contact area would decrease frictional forces developed (Abdel-Aal 2013). The geometry of the ridges represent sphere-sphere contact as opposed to the original flat surface-sphere contact, and according to the Hertzian theory of elastic mechanical contact, the contact area in sphere-sphere contact is half that of flat surface-sphere contact given the same radii of curvature. Therefore, the existence of the

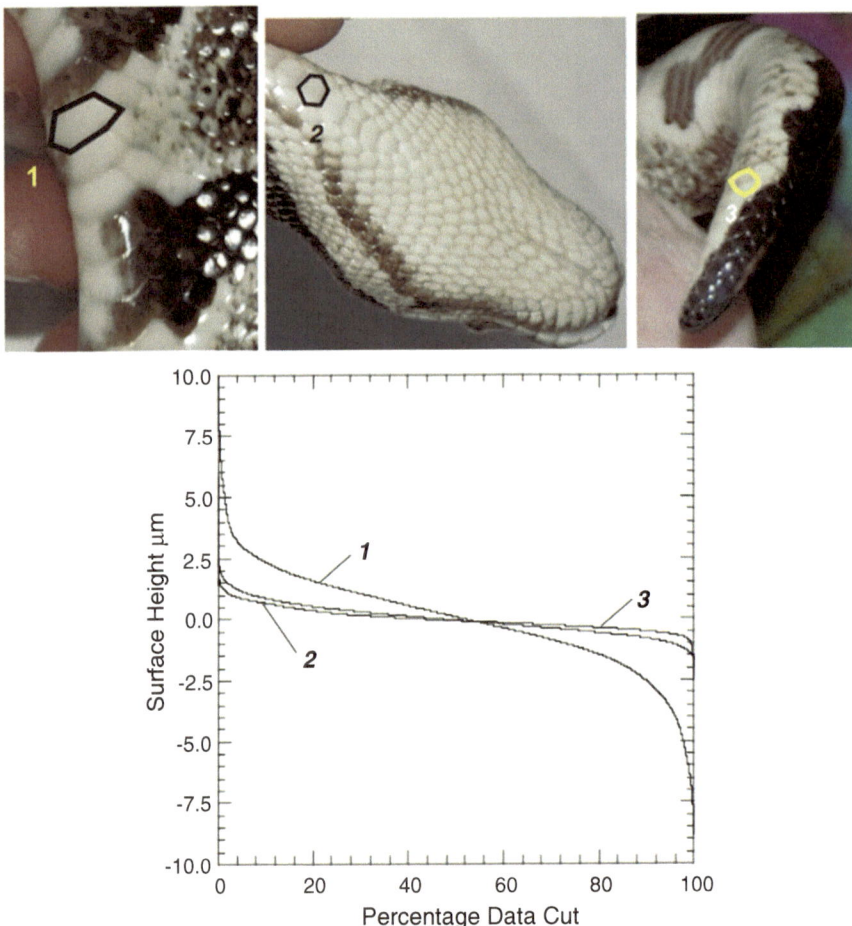

Fig. 9.6 Abbott-Firestone Load Curve (AFLC) of scale samples taken in regions as follows: middle trunk (*1*), neck-trunk (*2*), and tail (*3*) (Reprinted from Abdel-Aal and El Mansori (2011), Fig. 4.13. With kind permission from Springer Science + Business Media)

ridges halves the amount of adhesive forces and friction at play (Hazel et al. 1999). Though friction is necessary for traction in movement—as mentioned earlier, snakes can utilize friction to propel themselves forward—friction should generally be minimized, as it opposes movement (Abdel-Aal 2013).

Reduction of Friction Due to Micropits

Snake skin also features a system of evenly distributed micropits (see Fig. 9.7 top). These micropits range in size, measuring about 30–50 nm across, as well as concentration. Depending on the specific type of snake skin, micropits can be found in

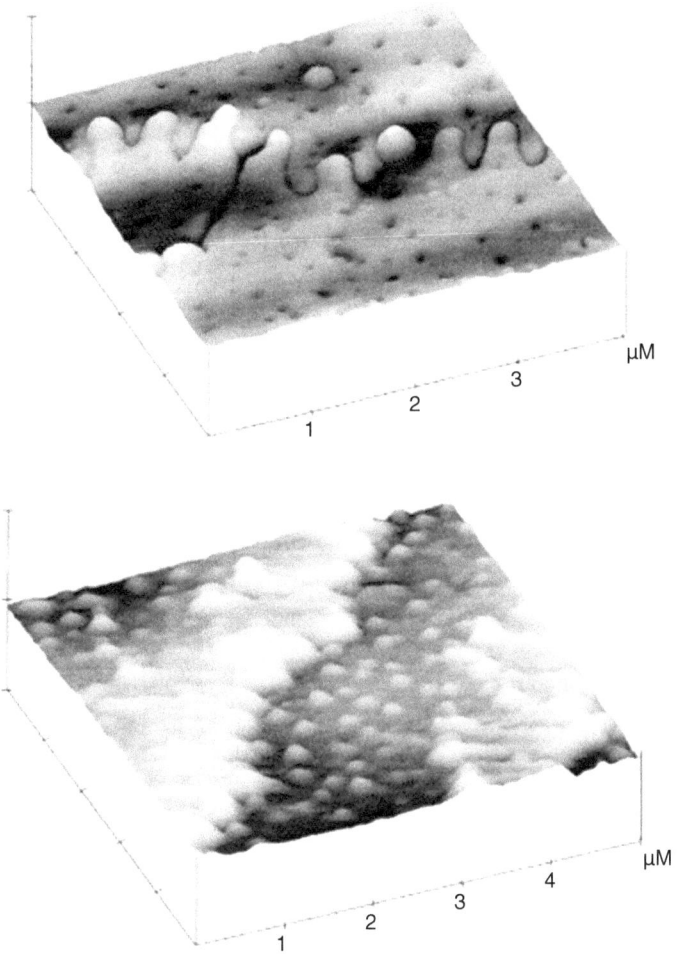

Fig. 9.7 *Top*: Micropit 3D topography of ball python skin (Z-range: 1 μm). *Bottom*: Hardened microdrops of viscous fluid after permeating through pore and curing (Z-range: 1 μm) (Reprinted from Hazel et al. (1999). With permission from Elsevier)

populations of 5–20 pits per square micrometer. After performing a microperme-ability test on the micropits, Hazel et al. were able to verify that they were actually pores through which fluid can permeate. The micropermeability test consisted of placing skin on top of viscous liquid and curing the result with slow-cure cyano acrylate glue. This process resulted in the confirmation of their hypothesis: almost every micropit was replaced with a small, hard bump, as seen in Fig. 9.7 bottom, which meant that the viscous liquid permeated through the pore and hardened at the top (Hazel et al. 1999).

These pores are present on dorsal scales of all snakes, indicating that they have a purpose common and critical to every species. Chiasson et al. suggest that the pores are the sole outlets for the secretion of epidermal lipids that act as the primary permeability barrier. They observed such a mechanism at play on the skin surface of *Nerodia erythrogaster*, where the esters of three fatty acids were found extruded through the pores (Chiasson et al. 1989). In addition to allowing the transmission of epidermal lipids that regulate permeability and wettability, the micropits are speculated to transmit lubricant to the surface of the skin in contact, thereby creating a boundary layer of organized molecular films and reducing friction forces up to ten times (Hazel et al. 1999).

Engineering Applications

Frictional Loss in Internal Combustion Engines (ICEs)

Due to snake skin's superior damage resistance, it has been identified as a potential analogue for the design of certain tribosystems. Applying the design features of snake skin to lubricating surfaces such as plateau honed surfaces is of particular interest to researchers, because the surface of snake skin is insightful for controlling wear and friction, as it has been shown in this chapter. Minimizing energy consumption, wear, and friction in the design of tribosystems is of great importance, especially since it has become a priority for engineers to reduce fossil fuel consumption (Abdel-Aal and El Mansori 2011).

Decreasing the energy consumption of internal combustion engines (ICEs) has been considered one of the main ways to achieve reduction in fossil fuel consumption. Given the large amount of energy consumed by friction—about 15 % of energy put into a vehicle is lost due to friction—reducing frictional energy loss can account for a significant portion of fuel savings (Priest and Taylor 2000). In ICEs, friction is generated from contact between metals or hydrodynamic stresses in metal and oil films, and it poses a particular problem between the cylinder and piston. With ICEs powering above 200 million vehicles in just the United States, more than a million barrels a day of crude oil is wasted on frictional loss.

The problem of friction between the cylinder and piston has much in common with the friction and wear demands on snake skin; that is, both systems are subject to large amounts of rubbing. Snake skin, with its natural ability to endure such high levels of abrasion, sliding, and rubbing, is naturally a source of inspiration for researchers aiming to design cylinder-piston systems for ICEs that are textured in a way to optimize efficiency and performance (Abdel-Aal and El Mansori 2011).

Plateau Honing and Snake Skin

Plateau honing is one such process that may optimize a cylinder and piston's sliding performance. Plateau honed surfaces are characterized by plateaus and grooves; plateaus are raised protrusions, while grooves are the valleys between the protrusions. In theory, this texture is supposed to reduce contact area between the metals of the piston and cylinder, thereby reducing contact friction, as well as retaining lubrication in the grooves for future replenishment. Snake skin has been observed to be a large-scale version of a plateau honed surface. In the snake's case, the hexagonal scales are the plateaus, and the soft tissue between the scales as boundaries are the grooves.

Though connectedness of surface texture is a major factor in lubrication quality and efficient use of lubrication, there has yet to be created a set of standardized values of such parameters for producing optimal performance of honed cylinders. The level of connectedness determines smoothness of oil flow and ensures retention of lubrication for smooth metal-to-metal sliding. Studies by Abdel-Aal and El Mansori of the geometry of the honed surface of snake skin at multiple scales have provided significant insight into a range of proportions that would ensure high-quality performance of the surface. After comparing Python skin with a plateau honed cylinder surface, Abdel-aal and El Mansori observed perfect connectedness of the grooves in Python skin and minimal variation of the geometrical proportions of the surface, confirming snake skin as a natural inspiration for honed surfaces capable of reducing fuel consumption (Abdel-Aal and El Mansori 2011).

Though snakes may be the number one feared animal in America, there is no doubt that they offer much more to us than just a scare. Possessing skin with remarkable wear resistance and fascinating microtopography, snakes have become a primary source of inspiration for applications in internal combustion engines and plateau honing. As engineers continue to explore snake skin further, we may be able to achieve optimization of efficiency and performance of select tribosystems at a level never reached before.

References

Abdel-Aal HA (2013) On surface structure and friction regulation in reptilian limbless locomotion. J Mech Behav Biomed Mater 22:115–135. doi:10.1016/j.jmbbm.2012.09.014

Abdel-Aal HA, El Mansori M (2009) Characterization of the frictional response of squamata shed skin in comparison to human skin. Paper presented at the 36th Leeds-Lyon symposium on tribology, Lyon, Sept 2009

Abdel-Aal HA, El Mansori M (2011) Reptilian skin as a biomimetic analogue for the design of deterministic tribosurfaces. In: Gruber P, Bruckner D, Hellmich C, Schmiedmayer H, Stachelberger H, Gebeshuber I (eds) Biomimetics—materials, structures, and processes. Springer, Heidelberg, pp 51–79. doi:10.1007/978-3-642-11934-7_4

Browne L (2012) Smooth snake scales. http://theforagingphotographer.wordpress.com/2012/09/18/smooth-snake-scales/. Accessed 4 Nov 2012

Chiasson RB, Bentley DL, Lowe CH (1989) Scale morphology in Agkistrodon and closely related crotaline genera. Herpetologica 45(4):430–438

Choset H (2010) Modular snake robots. http://biorobotics.ri.cmu.edu/projects/modsnake/index.html. Accessed 27 June 2013

Filippov A, Gorb SN (2013) Frictional-anisotropy-based systems in biology: structural diversity and numerical model. Sci Rep. doi:10.1038/srep01240

Hazel J, Stone M, Grace MS, Tsukruk VV (1999) Nanoscale design of snake skin for reptation locomotions via friction anisotropy. J Biomech 32(5):477–484. doi:10.1016/S0021-9290(99)00013-5

Hu DL, Nirody J, Scott T, Shelley MJ (2009) The mechanics of slithering locomotion. Proc Natl Acad Sci U S A 106(25):10081–10085. doi:10.1073/pnas.0812533106

Priest M, Taylor CM (2000) Automobile engine tribology—approaching the surface. Wear 241:193–203. doi:10.1016/S0043-1648(00)00375-6

Public Policy Polling (2013) Animals and pets poll: Americans prefer dogs; fear snakes. http://www.publicpolicypolling.com/main/2013/06/animals-and-pets-poll-american-prefer-dogs-fear-snakes.html. Accessed 29 Sept 2013

Spice B (2013) Carnegie Mellon snake robot winds its way through pipes, vessels of nuclear power plant. http://www.cmu.edu/news/stories/archives/2013/july/july9_snakerobot.html. Accessed 11 July 2013

Vogel A (2009) Slithering snakes: research shows snakes use friction and weight redistribution to glide on flat terrain. http://www.gtresearchnews.gatech.edu/snakes/. Accessed 11 July 2013

Michelle Lee

One Small Step for a Gecko, One Giant Leap for Dry Adhesion

Geckos are known for their ability to climb and stick to all kinds of surfaces in various orientations, be it a vertical tree or the ceiling of a room. Called smart or reversible adhesion, the way they adhere to surfaces features a dynamic attachment and detachment capability. However, other animals like lizards and frogs as well as many insects such as spiders, flies, and beetles are also capable of smart adhesion. Why, then, are scientists so intrigued with geckos, so much so that they are the most widely studied of all sticky-toed creatures? This can be answered with a two-part response. The first reason is that, of all adhesive animals and insects, geckos have the greatest body mass. Secondly, despite the larger load they must carry, geckos demonstrate a superior and more effective adhesion mechanism compared to that of their other sticky-toed friends (Bhushan 2012).

Gecko Adhesion

Early Theories

Gecko adhesion is strong, sufficient enough for geckos to scale vertical walls and stick to surfaces upside down. In fact, a gecko weighing about 40–50 g is capable of generating a force that is 100 times greater than its weight (Niewiarowski et al. 2008). One of the main features of gecko adhesion that scientists have been trying to replicate but still find elusive is its good performance over the course of many cycles of attachment and detachment (Lee et al. 2007). The mechanisms behind the

---

Particular gratitude to Elizabeth Bifano for allowing me to consult her draft on gecko pads.

M. Lee (✉)
Mechanical Engineering, McCormick School of Engineering
at Northwestern University, Evanston, IL 60208, USA
e-mail: MichelleLee2013@u.northwestern.edu

M. Lee (ed.), *Remarkable Natural Material Surfaces and Their Engineering Potential,*
DOI 10.1007/978-3-319-03125-5_10, © Springer International Publishing Switzerland 2014

phenomenal, sustained adhesion of gecko toe pads have been studied for a long time, with scientists attributing this property to a plethora of theories.

Previous to Autumn et al.'s research on gecko adhesion, the varying hypotheses for geckos' smart adhesion included sticky fluids, electrostatic attraction charges, frictional force, suction, and microinterlocking. Over time, however, these hypotheses began to be ruled out. For example, geckos' lack of glands that secrete the suggested sticky fluids eliminated that theory, while the lack of variation in attractive force under different pressures ruled out suction. The electrostatic attraction concept was eliminated based on the observation that geckos retained their adhesion even in ionized air in which there would be no electrostatic attraction charges. Microinterlocking was not possible as geckos adhered to surfaces without any roughness, which would be required for microinterlocking. Finally, since friction is a parallel force, the idea of frictional forces at play between gecko pads and surfaces was deemed incorrect (Bhushan 2012). In the end, thin-film capillary forces and van der Waals forces were the two competing theories for the mechanism behind geckos' dry adhesion. Though capillary forces—adhesion based on the hydrophilicity of the surface—and van der Waals forces—adhesion based on the shape and size of gecko setae—work in very different ways, verification of the gecko's primary adhesion mechanism had never been accomplished due to lack of testing (Autumn et al. 2002).

van der Waals Forces (and More)

To distinguish one or the other as the primary mechanism, Autumn et al. designed and executed an experiment in which the results would be dictated by whether or not the magnitude of the adhesion forces depended on the hydrophobicity and hydrophilicity of the surface. Tokay geckos (*Gekko gecko*) were the subjects of their experiments (Autumn et al. 2002). Male Tokay geckos can grow up to 0.4 m while females can grow up to 0.3 m, and they may weigh up to 300 g. These geckos are spotted orange or red with blue or gray bodies and are the most observed because of their size and wide availability (Bhushan 2012). The researchers tested the toes of nine Tokay geckos (the toes of a Tokay gecko can be seen in Fig. 10.1) on hydrophobic (Si and GaAs) and hydrophilic (SiO_2) semiconductor wafers and found that the toes generated high attachment forces on both the hydrophobic and hydrophilic surfaces. Based on these results, it was concluded that van der Waals forces are the primary determinant of gecko adhesion rather than capillary forces (Autumn et al. 2002). Considering the myriad environments in which geckos live, such as tropical forests, deserts, and even urban areas, van der Waals adhesion is optimal since it is not surface specific, rather depending on the weak intermolecular forces between the gecko's foot and the surface (Stark et al. 2012).

Though van der Waals is recognized as the main contributor to gecko adhesion, capillary forces may still be at play when water vapor exists. In such an environment, capillary forces can increase the adhesion force (Zhou et al. 2013). It has been found that in the case of hydrophilic surfaces, the level of humidity influences adhesive force. Researchers found that elevating relative humidity from 35 to 80 % generated

Fig. 10.1 Bottom of the foot of a Tokay gecko (*Gekko gecko*) (Photograph by David Clements)

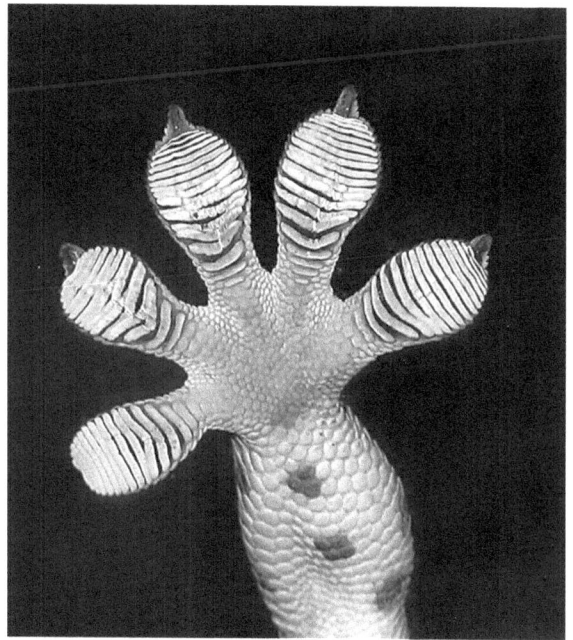

a 1.3-fold increase in adhesive force, and a 3-fold increase in adhesion occurred between relative humidities of 0 and 70 %. Although dependence on humidity is complex in itself and further complicated by gecko adhesion's dependence on temperature, such results do confirm the existence of capillary forces. It is also important to note that while experimental evidence has proven van der Waals forces to be the main mechanism behind gecko adhesion, the gecko as a whole should be taken into account, as vascular, skeletal, and muscular factors may also contribute to adhesion and detachment (Niewiarowski et al. 2008).

Hierarchical Morphology

Lamellae

At this point, it will not be surprising that geckos' adhesive ability comes from the hierarchical morphology of their toe pads, which can be seen in Fig. 10.2. So far, we have seen multiple examples of the prominence of hierarchical structure in nature, from strong nacre to the surfaces of moth eyes. Gecko pads also utilize the advantages of hierarchical structure, and in this case, it gives them the dynamic attaching and detaching skill they are so famous for (Bhushan 2012).

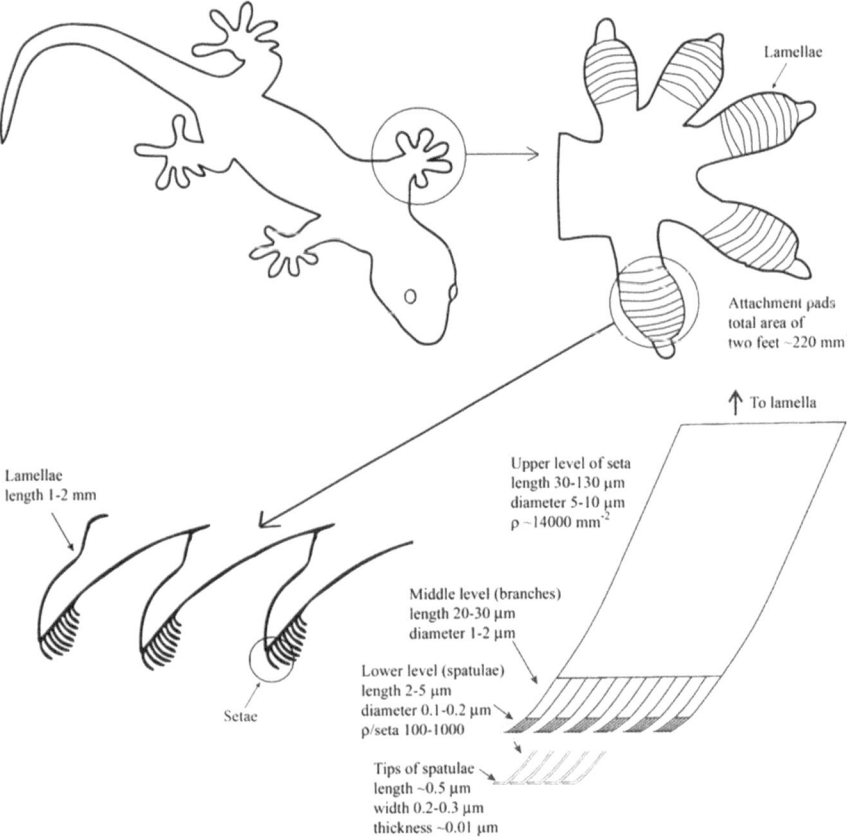

Fig. 10.2 Schematic of hierarchical structure of Tokay gecko feet: entire body, a single foot, a cross-sectional look at the lamellae, a single seta. ρ is the quantity of spatulae (Reprinted from Bhushan (2012), Fig. 3. With kind permission from Springer Science + Business Media)

The first level of hierarchy consists of lamellae, which are overlapping folds of skin on the digital pads of geckos (Ruibal and Ernst 1965). The folding creates ridges that are about 1–2 mm long, and their softness enables contact with non-smooth, rough surfaces (Bhushan 2012).

Setae and Spatulae

The lamellae are covered in microscopic setae (Fig. 10.3a, b), which look like hair-like projections. Setae are much thinner than hair, however, with its diameter ten times smaller than that of human hair (Autumn et al. 2000). Setae are made of some α-keratin, but β-keratin is the main component (Bhushan 2012). The setae are densely packed—researchers observed about one million setae per digital pad—and measure about 30–130 μm long, with the average being around 110 μm

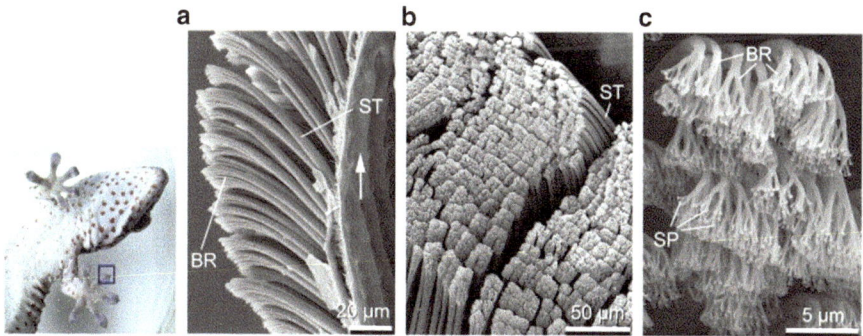

Fig. 10.3 Scanning electron micrographs show: (**a**) rows of setae (*ST*), (**b**) rows of setae (*ST*) at a differing magnification, and (**c**) setae (*ST*) branching (*BR*) into terminated ends called spatulae (*SP*) (Reprinted from Gao et al. (2005). With permission from Elsevier)

(Ruibal and Ernst 1965). The diameters of the setae range from 5 to 10 μm (Bhushan 2012), and each seta divides into 100–1,000 branches terminated by flat surfaces called spatulae, as shown in Fig. 10.3c (Zhou et al. 2013). The name given to these surfaces is not a coincidence and is based on their resemblance to the kitchen tool (Bhushan 2012).

Movement

Dry Adhesion

The force that allows geckos to stick to many surfaces and even walk upside down from glass ceilings is produced by setae. Though the total force is about 10 N cm-2, it is a cumulative effort resulting from millions of tiny forces—about 10–7 N per seta (Geim et al. 2003). In order to achieve such force, geckos' setae undergo a specific sequence of preloading and sliding that was observed by Autumn et al. These preloads consist of one in the perpendicular direction that is hypothesized to originate from the digital sinus system and another in the parallel direction thought to come from the lateral digital tendon system. The attachment process begins when contact is made with a surface, and the seta is subjected to preloading in the perpendicular direction. This is followed by parallel preloading that deflects the attachment angle between the surface and the seta so that it is under the critical angle of $30.6° \pm 1.8°$ due to tensile stress. The parallel preload also causes the tip of the seta to slide about 5 μm, at which point the maximum adhesive force is achieved and maintained until detachment. Based on the behavior of the gecko and the goal of its current motion, the time between attachment and detachment varies. It can be as short as milliseconds if, say, the gecko is running, and hours if the gecko intends to remain stationary (Russell 2002).

Geckos' dry adhesion mechanism can be thought of as a one-way adhesive according to Mark Cutkosky, professor of mechanical engineering at Stanford. Despite the gecko's ability to support its whole, hanging weight from the adhesion of one toe, pulling in a certain direction enables instant release (Blackman 2010). Setae exhibit friction anisotropy in which the elastic moduli of the arrays differ depending on whether the setae are sliding in the attaching or detaching directions. In the attaching direction, adhesion is increased as friction overcomes the preload by four times. Oppositely, the friction force decreases below the preload in the detaching direction, and repulsion occurs in the normal direction, releasing the gecko (Zhou et al. 2013).

Dry adhesion is made possible because of the surface structure of gecko pads. The splitting of setae into hundreds and even up to a thousand spatulae enhances adhesion and is known as the "contact-splitting principle". The more splitting of the setae, the more adhesive force is generated (Zhou et al. 2013). Furthermore, the lamellae of gecko pads have an important role in the control of adhesion. Due to the flexibility and softness of lamellar skin, it can be seen as a spring with a weak spring constant as opposed to stiff skin. This weak springiness ensures adhesion to surfaces over a wider range of normal compression displacement by preventing large deformations among the setae. If the skin were stiff, great repulsive forces would be produced as the crowded populations of setae undergo compression during deformation (Tian et al. 2013). Additionally, the cushioning effect of the lamellae encourages and increases the potential for each seta to come into contact with the surface, which in turn increases the adhesion forces generated since they must occur at interfaces of contact (Russell 2002).

Release

In addition to providing a formidable adhesive force, setae also enable effective detachment through the peel mechanism. The peel mechanism starts when the gecko uncurls its toes (Bhushan 2012). When the attachment angle between the surface and a single seta approaches a critical angle of $30.6° \pm 1.8°$ as the toe is uncurled, the seta is then released. Rows of setae detaching successively in this way ultimately lead to the overall peeling action. A benefit of the toe peeling process is that the detachment force is concentrated on just one portion of setae at a time (Autumn et al. 2000).

Engineering Applications

Adhesives Inspired by Geckos

Adhesives based on the morphology of gecko pads have been explored with much depth for quite some time due to their wide applicability. Successful gecko-inspired adhesives would boast an array of outstanding properties, such as the ability to

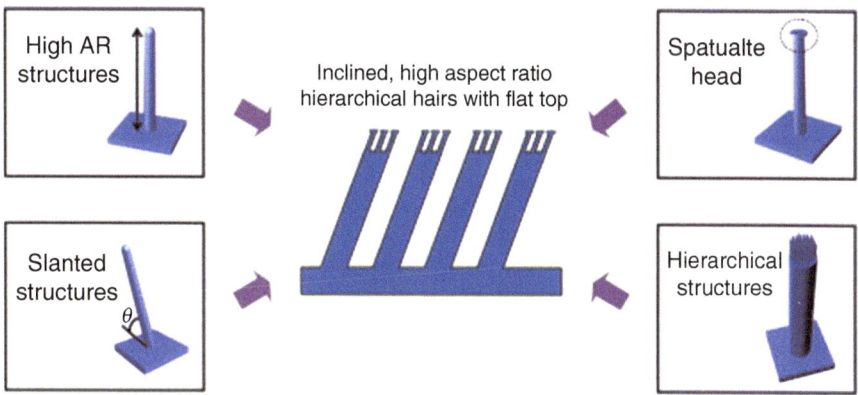

Fig. 10.4 Superior dry adhesives require four structural features, shown in the schematic above (Reprinted from Jeong and Suh (2009). With permission from Elsevier)

adhere and detach repetitively without breakage and remarkable strength coupled with easy movement. Uses include sports equipment and robots that climb walls, among a variety of others (Zhou et al. 2013).

For the fabrication of well-performing artificial adhesives, however, there are several structural requirements. According to Jeong et al., the requirements can be organized into four features that are as follows: (a) nanostructures with high aspect ratios, (b) slanted structures, (c) structures terminated in spatulate heads, and (d) hierarchical structures. These four features can be seen in Fig. 10.4.

The necessity of an artificial adhesive's nanostructures to have high aspect ratios comes from the principle that increased pillar amount, decreased effective modulus, and increased dissipation of elastic energy upon pull-off enhances adhesive force. Furthermore, slanted structures provide the anisotropic property that enables geckos to adhere with strength but detach with ease. If these slanted, high aspect ratio nanostructures terminate in spatulate heads as opposed to spherical heads, contact area would be increased and adhesion further enhanced (see Fig. 10.5). Finally, the production of hierarchical structures would contribute to better adaptability on non-smooth surfaces.

With these requirements in place, two general fabrication methods—polymer-based and carbon nanotube (CNT)-based dry adhesives—have been developed to create gecko-inspired adhesives. Though each method boasts their own list of advantages and suffers from their respective disadvantages, the overall agreement is that polymer-based adhesives are precision-driven, simple, economic, and quick to produce, while CNT-based adhesives exhibit stronger adhesion and better mechanical and structural properties. The existence of two very different types of dry adhesives can be seen as an advantage, because their respective superior properties can be taken advantage of and applied to suitable applications. For instance, polymer-based adhesives are good for biomedical applications due to their high precision and processability, and CNT-based adhesives may be fabricated for uses that demand strength and durability (Jeong and Suh 2009).

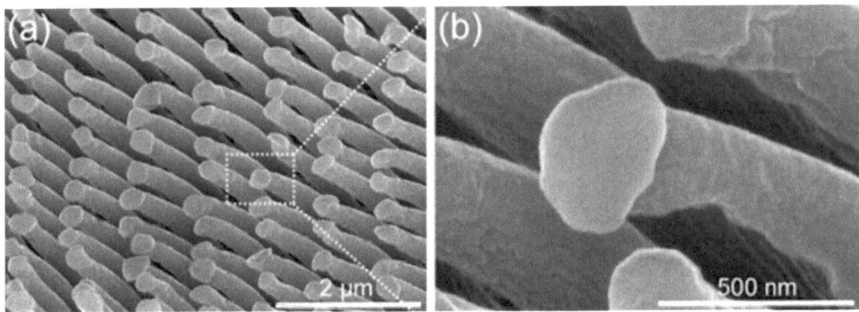

Fig. 10.5 Dry adhesives being developed today terminate in mushroom-like heads like spatulae due to the superior adhersive power they produce. Artificial dry adhesives with such spatulae-like ends have been proven to generate adhesive forces 3–30 times greater than their non-spatulae-like counterparts. (**a**) Bulged, flat tips of polyurethane acrylate nanohairs in SEM image and (**b**) at higher magnification (Reprinted from Jeong and Suh (2009). With permission from Elsevier)

Polymer-Based Adhesives: Prostheses in the Biomedical Arena

Application in the realm of prostheses is one example of potentially using polymer-based adhesives in biomedicine. Prostheses are usually held in place with adhesive aid, and an example of such an application would be maxillofacial prostheses, which are used by patients suffering from head and neck defects. Such prosthetic adhesives would be made of nanofibers that exhibit the gecko effect and are attractive options since they meet all the requirements of an ideal adhesive for prostheses. That is, they are low maintenance and easy to handle and clean, they are strong yet easy to remove when needed, and if made from biocompatible materials, they would not cause irritation to the patient.

To explore the viability of gecko-inspired adhesives for prosthetic applications, Palacio et al. fabricated single-level and hierarchical surfaces through the membrane template technique. The membrane templating technique consists of forcing a polymer through pores by applying pressure and heat and dissolving the membrane with a solvent that would not damage the nanofibers. Some advantages of this process are its relative cheapness and simplicity. Despite previous attempts to create a single-level, nanofiber surface out of polysiloxane, the resulting nanofibers were limp and collapsed, leading Palacio et al. to use polyethylene in their experiment, as polyethylene has enhanced mechanical properties.

Three polyethylene surfaces were observed in this experiment: a flat surface, a single-level surface (Fig. 10.6, top), and a hierarchical surface (Fig. 10.6, bottom)—the last two of which were fabricated using membrane templating. Following fabrication, researchers tested the adhesive forces of the nanofiber surfaces using atomic force microscopy (AFM) and made two observations that led them to confirm hierarchical structures as promoters of gecko adhesion. First, the amount of adhesive force generated by the nanofiber surfaces were both larger than that generated by the flat surface, and secondly, the hierarchically structured surface generated the

Single-level fibers (0.6 µm)

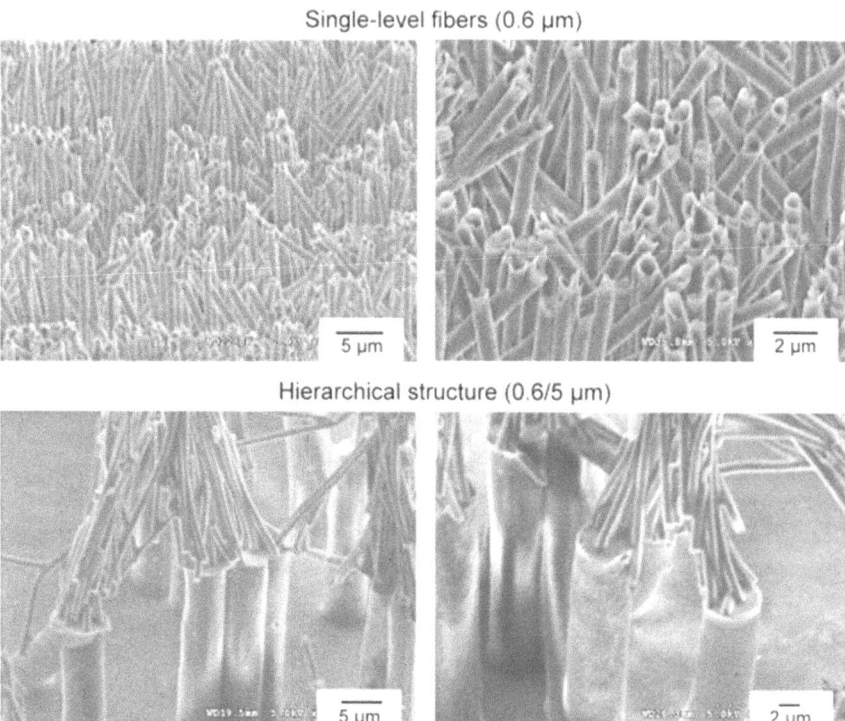

Hierarchical structure (0.6/5 µm)

Fig. 10.6 (**a**) Single-level and (**b**) hierarchical nanofiber structure images by scanning electron microscopy (SEM). Fibers per unit area are an estimated 0.6 and 0.03 µm^{-2} for (**a**) and (**b**), respectively (Reprinted from Palacio et al. (2013). With permission from Elsevier)

highest adhesion. Palacio et al.'s success in creating the first polyethylene single- and hierarchical-level surfaces via membrane templating makes them attractive candidates for adhesion in biomedical applications. Though utilization for applications like maxillofacial prostheses would require additional research, this is a strong step towards superior adhesives for these important applications (Palacio et al. 2013).

Carbon Nanotube-Based Adhesives: Major Sticking Power

Teams of researchers at Daytona University and Georgia Institute of Technology led by Liming Dai and Zhong Lin Wang, respectively, fabricated carbon nanotube-based adhesives that outperforms the strong grip of a gecko by ten times and is three times stronger than their top competitor in artificial adhesives. Their carbon nanotube-based adhesive is able to withstand 100 N in the shear direction, even though its area measures only 1 cm^2. However, just like a gecko pad, the adhesive is easy to remove when tugged in a certain direction. To create this superb adhesion strength, the team of researchers grew 4 by 4 mm arrays of carbon nanotubes in a way that the ends tangled with each other. The resulting structures resemble those of a gecko's setae and spatulae: the main trunk is like a seta, and the coiled end is like a spatula (Griggs 2008).

Future Outlook

Recent research has brought to light another mechanism at play in gecko adhesion that could change the way scientists approach the fabrication of gecko-inspired artificial adhesives. In 2012, Hsu et al. utilized nano-assisted laser desorption mass spectrometry (NALDI-MS) to identify the footprint-shaped residues left on surfaces just walked upon by geckos. Their findings revealed the residue to be phospholipids with phosphocholine head groups, and chemical mapping showed that the residues appeared at locations where the hairs of the gecko pad touched the surface. The researchers believe that the lipid molecules may potentially contribute to geckos' ability to control the peeling of their toes during movement, as well as protect their setae and spatulae from degradation and wear. Overall, an important implication of Hsu et al.'s findings is that researchers may want to incorporate lipid layers in their models of gecko adhesion and their fabricating processes. The addition of such layers may be an important factor in producing adhesives that last (Hsu et al. 2012).

A problem that researchers have been having with gecko-inspired artificial adhesives is the inability to scale the nano-scale features up to perform well at large scales. As the surface area of a sample increases, it becomes less probable to create the intimate contact required for van der Waals forces to be generated. Furthermore, compared to the projected contact area, the real contact area is much smaller—usually a mere fraction.

To address this issue, Bartlett et al. used a simple scaling framework to create a macro-scale gecko-inspired adhesive with unprecedented performance. Using an energy balance principle, Bartlett et al. found the scaling proportion—regardless of geometry, size, and shape—for the maximum sustainable force, F_C, to be:

$$F_c \sim \sqrt{G_c}\sqrt{\frac{A}{C}}$$

where G_C is a property dependent upon the materials of the interface, A is the surface area, and C is the system compliance. Just as ancient Greeks were unsuccessful in their attempts to fly using keratinous features like feathers, Bartlett et al. believe that replicating form alone is, in many cases, not capable of producing outcomes usable for desired applications today. This scaling model addresses the demands of specific applications—that is, larger scale—by employing the functional principles behind gecko adhesion. Understanding that a larger scale reversible adhesive would need the ability to conform but be stiff in the loading direction, the researchers used elastomer fabric that, on the small scale, maximizes contact through flexibility, yet on the large scale, causes the necessary stiffness. The result? A reversible, artificial adhesive the size of a hand that can hold up to 2950 N (or about 660 lb) (Bartlett et al. 2012). This success not only has great implications for the future of synthetic adhesives at large scales, but it is also a great example for achieving outstanding performance through bioinspiration at the functional level.

Gecko pads utilize an abundance of characteristics and techniques to achieve the strong attachment and dynamic detachment they are so well known for. These include the multi-level hierarchical structure of the surface of their toe pads, their dry adhesion technique, their peel mechanism, and more. Whether it is polymer-based or carbon nanotube-based adhesives, gecko pads are helping pave the way for engineers in the realm of artificial adhesives. Outstanding examples of biomimicry in engineering applications, gecko-inspired adhesives demonstrate the possibility of developing powerful solutions to current problems through the exploration of natural materials.

References

Autumn K, Liang YA, Hsieh ST, Zesch W, Chan WP, Kenny TW, Fearing R, Full RJ (2000) Adhesive force of a single gecko foot-hair. Nature 405:681–685. doi:10.1038/35015073

Autumn K, Sitti M, Liang YA, Peattie AM, Hansen WR, Sponberg S, Kenny TW, Fearing R, Israelachvili JN, Full RJ (2002) Evidence for van der Waals adhesion in gecko setae. Proc Natl Acad Sci U S A 99(19):12252–12256. doi:10.1073/pnas.192252799

Bartlett MD, Croll AB, King DR, Paret BM, Irschick DJ, Crosby AJ (2012) Looking beyond fibrillar features to scale gecko-like adhesion. Adv Mater 24(8):1078–1083. doi:10.1002/adma.201104191

Bhushan B (2012) Gecko effect. In: Bhushan B (ed) Encyclopedia of nanotechnology. Springer, Netherlands, pp 943–951. doi:10.1007/978-90-481-9751-4_378

Blackman C (2010) Secrets of the gecko foot help robot climb. http://news.stanford.edu/news/2010/august/gecko-082410.html. Accessed 22 Aug 2013

Gao H, Wang X, Yao H, Gorb S, Arzt E (2005) Mechanics of hierarchical adhesion structures of geckos. Mech Mater 37(2–3):275–285. doi:10.1016/j.mechmat.2004.03.008

Geim AK, Dubonos SV, Grigorieva IV, Novoselov KS, Zhukov AA, Shapoval SY (2003) Microfabricated adhesive mimicking gecko foot-hair. Nat Mater 2:461–463. doi:10.1038/nmat917

Griggs J (2008) Gecko sticking power outclassed by nanotubes. New Sci 200(2678):26. doi:10.1016/S0262-4079(08)62630-5

Hsu PY, Ge L, Li X, Stark AY, Wesdemiotis C, Niewiarowski PH, Dhinojwala A (2012) Direct evidence of phospholipids in gecko footprints and spatula-substrate contact interface detected using surface-sensitive spectroscopy. J R Soc Interface 9(69):657–664. doi:10.1098/rsif.2011.0370

Jeong HE, Suh KY (2009) Nanohairs and nanotubes: efficient structural elements for gecko-inspired artificial dry adhesives. Nano Today 4(4):335–346. doi:10.1016/j.nantod.2009.06.004

Lee H, Lee BP, Messersmith PB (2007) A reversible wet/dry adhesive inspired by mussels and geckos. Nature 448:338–342. doi:10.1038/nature05968

Niewiarowski PH, Lopez S, Ge L, Hagan E, Dhinojwala A (2008) Sticky gecko feet: the role of temperature and humidity. PLoS One 3(5):e2192. doi:10.1371/journal.pone.0002192

Palacio MLB, Bhushan B, Schricker SR (2013) Gecko-inspired fibril nanostructures for reversible adhesion in biomedical applications. Mater Lett 92:409–412. doi:10.1016/j.matlet.2012.11.023

Ruibal R, Ernst V (1965) The structure of the digital setae of lizards. J Morphol 117(3):271–293. doi:10.1002/jmor.1051170302

Russell AP (2002) Integrative functional morphology of the Gekkotan adhesive system (Reptilia: Gekkota). Integr Comp Biol 42(6):1154–1163. doi:10.1093/icb/42.6.1154

Stark AY, Sullivan TW, Niewiarowski PH (2012) The effect of surface water and wetting on gecko adhesion. J Exp Biol 215:3080–3086. doi:10.1242/jeb.070912

Tian Y, Wan J, Pesika N, Zhou M (2013) Bridging nanocontacts to macroscale gecko adhesion by sliding soft lamellar skin supported setal array. Sci Rep 3:1382. doi:10.1038/srep01382

Zhou M, Pesika N, Zeng H, Tian Y, Israelachvili J (2013) Recent advances in gecko adhesion and friction mechanisms and development of gecko-inspired dry adhesive surfaces. Friction 1(2):114–129. doi:10.1007/s40544-013-0011-5

Shiqi Luohong

The Colorful World of Butterflies

Butterflies are, without a doubt, some of the most beautiful insects that can be found in nature. One glance upon a butterfly lightly skirting a flower on a spring day is enough to convince many that its elegance, color, and beauty are unrivaled in the insect world. Watching a butterfly flutter across a lush green meadow or freshly mown lawn is sure to lift the spirits of someone having the most down of days. Their thin wings display bold, kaleidoscope-like patterns with a rainbow of vivid colors and artfully arranged shapes. However, there is even more to butterfly wings than the delight they bring to our eyes. They are natural surfaces that, when looked upon closely, offer a plethora of scientific knowledge and biomimetic inspiration.

Butterfly Wing Morphology

Inspirational Structures

The butterfly amazes people not just for its beautiful, kaleidoscope-like appearance but also its potential for mimicry, which is evident from its high adaptability in various kinds of environments. There are nearly 18,000 kinds of butterflies in the world (Hoskins 2013). Each has a particular wing structure and corresponding living habit. Although their iridescent colors attract the most attention, other characteristics like their anti-wetting property and heat dissipation due to their special surface structure are also interesting and inspiring.

S. Luohong (✉)
Mechanical Engineering, McCormick School of Engineering
at Northwestern University, Evanston, IL 60208, USA
e-mail: ShiqiLuohong2012@u.northwestern.edu

M. Lee (ed.), *Remarkable Natural Material Surfaces and Their Engineering Potential,*   127
DOI 10.1007/978-3-319-03125-5_11, © Springer International Publishing Switzerland 2014

Fig. 11.1 Forewing and hindwing of *Parantica sita* (Reprinted from Kang et al. (2012). With permission from Elsevier)

General Morphology

Prior to investigating the scientific nature of butterfly wings, it is important to understand their anatomy. A butterfly always has four wings: two fore-wings and two hind-wings (see Fig. 11.1). Both pairs are symmetrical. Also, we define the dorsal and ventral side of butterfly wings by their different surface characteristics and purposes in their lives (Arkian 2011). The surface of the dorsal side of the butterfly is always covered in pigment beads that contribute to the wing's brilliant appearance, while the ventral side has a lower reflectance of light (Stavenga 2009).

Micro-scale Structure

On the micro-scale, we see that the wings of butterflies are not as smooth as they seem. Zheng et al. investigated the wings of *Morpho aega* through scanning electron microscopy (SEM) and atomic force microscopy (AFM). SEM observations (see Fig. 11.2a) show that the wing surface is covered by a large number of quadrate scales with a length of ~150 μm and a width of ~70 μm. They overlap each other and are orderly arranged along the radial outward (RO) direction (Zheng et al. 2007).

Nano-scale Structure

When the image was magnified to the nano-scale level, research revealed that each scale consisted of ridging stripes, 184.3 ± 9.1 nm in width and 585.5 ± 16.3 nm in clearance (see Fig. 11.2b). These nano-stripes can be further split into multiple layers separated by air, which are cuticle lamellae of different lengths. They stack stepwise along the RO direction. Observation shows that the tops of the stripes (nano-tips) are tilted slightly upward (see Fig. 11.2b). AFM scanning provides more detailed information about the structural parameters, particularly the overlapping scales, which are flexibly fluctuated with a peak height of ~6 μm, as seen in Fig. 11.2c, while the nano-tips are tilted with a peak height of 121.3 ± 21.7 nm, as seen in Fig. 11.2d (Zheng et al. 2007).

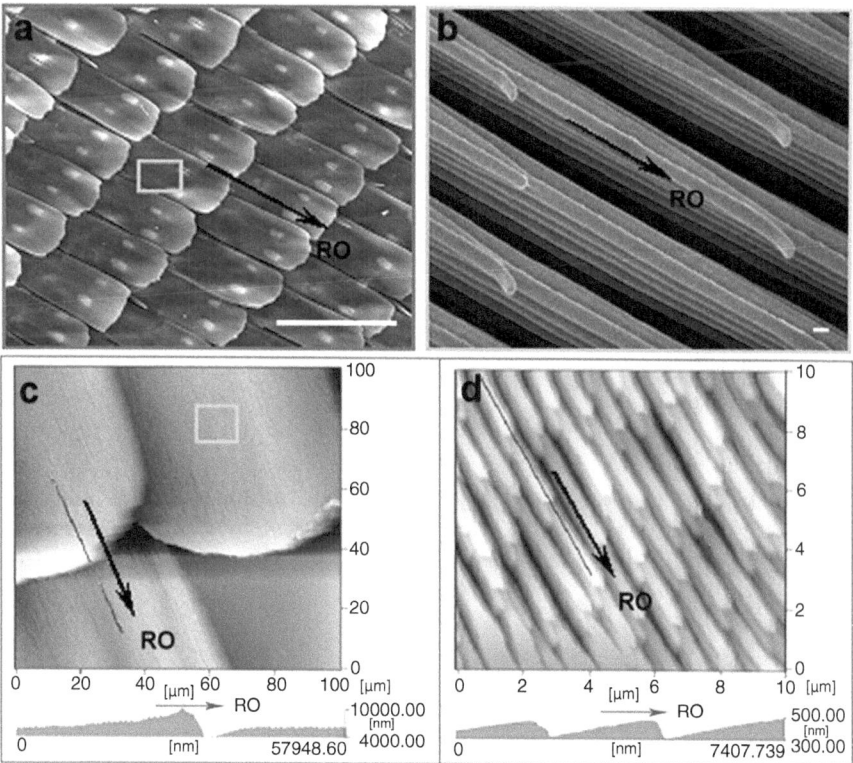

Fig. 11.2 Multi-layered nano- and micro-structures of a butterfly wing surface: (**a**) scanning electron microscopy (SEM) images of the arrangement of overlapping micro-scales on the wing, (**b**) SEM images of fine lamellae-stacking nano-stripes on the scales, (**c**) atomic force microscopy (AFM) images of the overlapping micro-scale structure, and (**d**) AFM images of nano-stripe structure. 'RO' indicates radial outward direction. Scale bars of (**a**) and (**b**) are 100 mm and 100 nm, respectively. The *black lines* of (**c**) and (**d**) represent the respective cross-sections (Reproduced from Zheng et al. (2007). With permission of The Royal Society of Chemistry)

However, the scales of some butterfly species display different structures. For example, they can be thin-film or three-dimensional periodical structures called photonic crystals (Stavenga 2009). Other kinds of butterflies may have irregularly arranged scales.

Scientific Nature of Butterfly Wings

Iridescence

The bright colors of butterfly wings are closely related to the micro- and nano-structures on the surface of the scales (Jiang et al. 2012). Scattering, which can be categorized into coherent and incoherent, is the primary mechanism property for animal coloration. Coherent scattering occurs when the surface layer of scales have dimensional periodicities.

According to Herring, the iridescent effect observed in many species is the product of the periodical multi-layers of their scales, which give rise to multiple internal reflection, refraction, and interference events. Butterflies with this typical wing structure can display a variety of common optical phenomena: reflection, refraction, and interference, as mentioned previously, but also fluorescence, iridescence, and so on. For instance, the blue iridescent effect of the *Morpho* butterfly is the result of its periodical multi-layer scales, which enable the wings to scatter coherent light (Herring 1994).

However, for butterflies with an irregularly-spaced multi-layer structure, light scattering would be incoherent and generally would not cause a distinct color. Their colors are caused by spectrally selective, absorbing pigments incorporated inside the scales (Goodwyn et al. 2009). The African Swallowtail butterfly, for example, contains a pigment that can obtain the ultraviolet (UV) component from the incident light and reserve the fluorescence in the micro-holes of their wings to prevent it from loss. In this way, the intensity of the emitted light is also enhanced (Kang et al. 2010). The unique structure of the surface of butterfly wings enables this creature to appear bright and colorful for signaling or camouflaging for protective purposes.

Superhydrophobicity

How do butterflies keep their wings clean and dry? Many research results demonstrate that it is due to the directional superhydrophobicity of their wings. Like lotus leaves and rice leaves, most butterfly wings are considered to be superhydrophobic, meaning that the contact angle with a water drop on its surface exceeds 150°. When a droplet is placed on the wing's surface, it can easily roll off along the radial outward (RO) direction of the central axis of the body or be pinned tightly against the RO direction. Butterflies are able to tune these two different states by controlling its posture (upward and downward) (Zheng et al. 2007).

Observations show that air pockets in the rough micro-structures may effectively reduce the area of water in contact with the surface of the wings, so it results in the nearly spherical droplet. As for the directional difference, Zheng et al. gave two reasonable hypothetical modes to clarify the distinct adhesive properties of butterfly wings. As shown in Fig. 11.3a, when the wing is tilted downward, the micro-scales with ridged nano-stripes are spatially separated from each other; thus, the nano-tips tend to be unwound with each other. As a result, air is efficiently trapped in these nano-scale voids among the nano-tips, allowing the droplet to touch only the top of the nano-tips. In this case, the wing is superhydrophobic with a high contact angle, and the droplet can easily roll off along the RO direction due to the discontinuous three-phase contact line (TCL) and orderly arranged micro-scales, as illustrated in the figure. As the droplet contacts the nano-tips and air in this rolling state, it can be treated as the Cassie State, which was introduced in earlier chapters and is a concept widely used in describing superhydrophobic states with extremely low adhesion.

Figure 11.3b shows the situation in which the wing is tilted upward. The nano-tips on the top of the nano-stripes are raised with the flexible micro-scales and thus

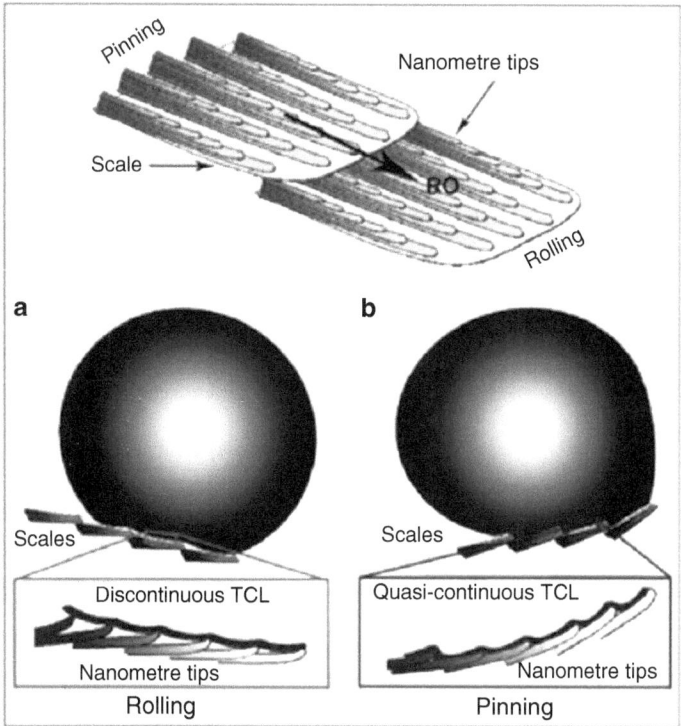

Fig. 11.3 The models for distinct adhesive properties on the direction (**a**) along and (**b**) against the radial outward (RO) direction (Reproduced from Zheng et al. (2007). With permission of The Royal Society of Chemistry)

closely contact the droplet. As a consequence, the solid-liquid contact area is increased while the air trapped in the nano-grooves stays constant. Since a quasi-continuous TCL is also formed in the process, the pinning at numerous corners (nano-tips) of the steps between the neighboring lamellae on top of the ridged nano-stripes produces a very high energy barrier, which makes the droplet pin tightly on the wing as it is tilted upward. Zheng et al. believe this to be an intermediate state with the "wet" contact on the nano-tips and the "dry" contact on the air trapped in the nano-grooves (Zheng et al. 2007).

Engineering Applications

Butterfly Wing Replication

By learning about the scientific nature of the surface of butterfly wings, we can mimic this surface to achieve materials that exhibit these outstanding properties. For example, mimicking the iridescent property of butterfly wings can help in the design

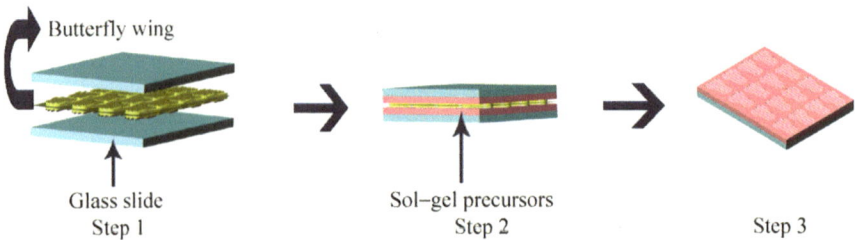

Fig. 11.4 Fabrication process of the SiO₂ inverse structure replica of butterfly wing scales. Step 1: A slice of butterfly wing is sandwiched between two glass slides. Step 2: Silica precursors are added to the edge of the assembly and heated at 100 °C for 20 min. Step 3: Wing is removed with acid-etching, and silicon inverse structure is washed with deionized water (Reproduced from Xu et al. (2011), Fig. 1, published by Tsinghua Press. With kind permission from Springer)

of paints and light. Furthermore, the hydrophobic property can be applied in the design of swimming suits, waterproof paint, and many other objects that would need to repel water. Many scientists have devoted much research to this field and have made significant progress in the replication of the butterfly-wing surface concept.

Dai et al. have proposed a procedural texture generation approach using traits of iteration behavior to achieve the surface texture of butterfly wings. They built a fundamental pattern by modeling the pattern of eye-like spots of select attractive fixed points and placing river like bands between them. Then, they produced variations of the fundamental pattern by changing parameters, thus creating different patterns of fore-wings and hind-wings. This method works well in constructing the texture pattern and colorful variations of butterfly wings (Dai et al. 1995).

The outstanding optical properties of butterfly wings also attract the attention of numerous scientists who aim to enhance the performance of optical devices. Xu et al. produced their SiO_2 inverse structure replicas using butterfly wings as templates in a sol-gel process, as shown in Fig. 11.4. First, they sandwich a slice of butterfly wing between two glasses, and then the silica precursor is added to the edge of the assembly. After being heated at 100 °C for 20 min, the wing is removed by acid-etching, and the silicon structure is washed with deionized water. Investigation by SEM, as shown in Fig. 11.5, reveals that the replica succeeded in copying the morphology of the wing surface but failed in duplicating the overlapping scales (Xu et al. 2011).

Another replication was created by Kang et al. using a molding lithography technique to fabricate a polydimethylsiloxane (PDMS) replica of the multi-layered scales on the upper surface of a *Morpho* butterfly, as seen in Fig. 11.6. The results show that the micro-structural and optical characteristics of the replicated wing agree with those of the actual wing. Furthermore, the contact angle for the natural wing structure and the replicated wing were about 143° and 120° respectively, resulting in a replica that displayed hydrophobic behavior, even with the slightly smaller contact angle (Kang et al. 2010).

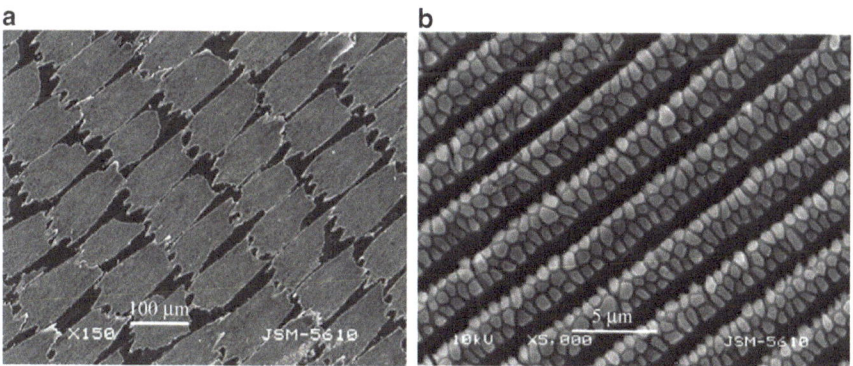

Fig. 11.5 SEM images of the SiO_2 inverse structure replica in different scales of magnification: scale bars (**a**) 100 µm and (**b**) 5 µm (Reproduced from Xu et al. (2011), Fig. 1, published by Tsinghua Press. With kind permission from Springer)

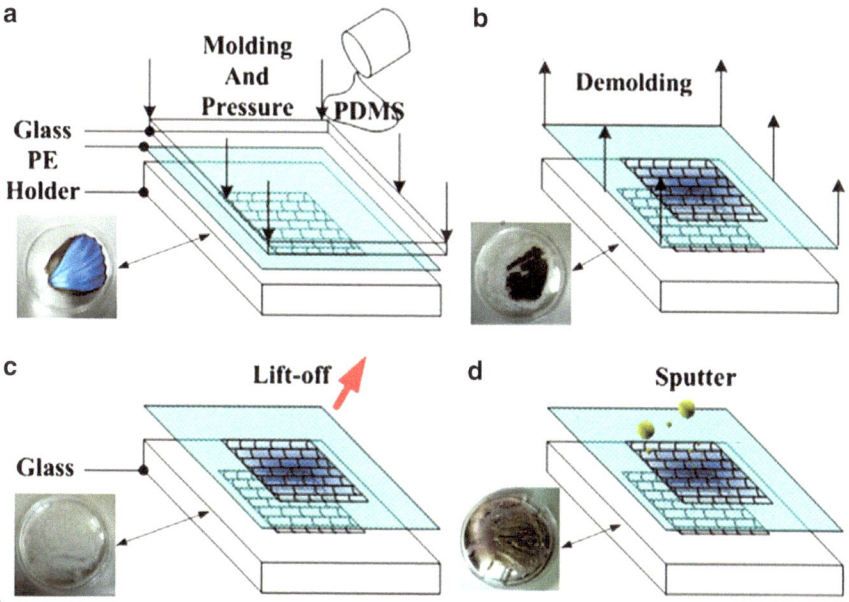

Fig. 11.6 PDMS-based soft lithography technique for replication of the *Morpho* butterfly wing: (**a**) PDMS (mixed with hardening agent) was poured into a mold containing a butterfly wing. The mold was placed under a polyethylene (PE) plate and a glass substrate with pressure. The mold was transferred and heated in a furnace until the PDMS solidified; (**b**) PDMS replica and PE plate was separated from the mold and placed on another glass substrate; (**c**) lifted off the PE plate, leaving only the PDMS replica on the glass substrate; (**d**) the substrate was placed in a vacuum chamber, and PDMS wing structure was sputtered with a thin layer of Pt/Au through a plasma deposition process to copy the pigmentation of the *Morpho* butterfly wing (Reprinted from Kang et al. (2010). With permission from Elsevier)

Prospective Trends

The major characteristics that contribute to the high optical and hydrophobic property of butterfly wings are the micro- and nano-scale structures that shingle the wings. In order to mimic this structure, replication work has been done to acquire the outstanding quality of butterfly wings through engineering methods. However, the fabrication of butterfly-wing-mimicking material is still not widely applicable in industry due to the complicated procedures, weak strength, and expensive apparatuses required by the manufacturing process. Furthermore, the existence of micro- and nano-structures leads to a weakness in mechanical strength (Shi and Feng 2012). In addition, the influence of the diameters of nano-granules and the spacing size of the gap between them should be investigated more. Therefore, building a theoretical model that fits well with the actual surface and optimizing the parameters that affect the material's performance are the most important topics to address in the future. Modeling, analysis, and exploration of creative fabrication techniques are critical in moving forward with these topics.

References

Arkian M (2011) Temperature control mechanism of butterfly wings. 3rd year mechanical report, University of Manchester

Dai WK, Chang RC, Shih ZC (1995) Fractal pattern for a butterfly wing. Visual Comput 11(4): 177–187. doi:10.1007/BF01901514

Goodwyn PP, Maezono Y, Hosoda N, Fujisaki K (2009) Waterproof and translucent wings at the same time: problems and solutions in butterflies. Naturwissenschaften 96(7):781–787. doi:10.1007/s00114-009-0531-z

Herring PJ (1994) Reflective systems in aquatic animals. Comp Biochem Physiol A Physiol 109(3):513–546. doi:10.1016/0300-9629(94)90192-9

Hoskins A (2013) Learn about butterflies: the complete guide to the world of butterflies and moths. http://www.learnaboutbutterflies.com/index.htm. Accessed 11 Nov 2012

Jiang X, Shi TL, Zuo HB, Yang XF, Wu WJ, Liao GL (2012) Investigation on color variation of Morpho butterfly wings hierarchical structure based on PCA. Sci China Technol Sci 55(1): 16–21. doi:10.1007/s11431-011-4528-4

Kang SH, Tai TY, Fang TH (2010) Replication of butterfly wing microstructures using molding lithography. Curr Appl Phys 10(2):625–630. doi:10.1016/j.cap.2009.08.007

Kang SH, Song SH, Lee SH (2012) Identification of butterfly species with a single neural network system. J Asia Pacific Entomol 15(3):431–435. doi:10.1016/j.aspen.2012.03.006

Shi Y, Feng X (2012) Progress in superhydrophobic bio-surfaces. Chin J Appl Chem 29(5): 489–497. doi:10.3724/SP.J.1095.2012.00328

Stavenga DG (2009) Surface colors of insects: wings and eyes. In: Gorb SN (ed) Functional surfaces in biology: little structures with big effects, vol 1. Springer, Netherlands, pp 285–306. doi:10.1007/978-1-4020-6697-9_15

Xu Z, Yu K, Li B, Huang R, Wu P, Mao H, Liao N, Zhu Z (2011) Optical properties of SiO_2 and ZnO nanostructured replicas of butterfly wing scales. Nano Res 4(8):737–745. doi:10.1007/s12274-011-0130-0

Zheng Y, Gao X, Jiang L (2007) Directional adhesion of superhydrophobic butterfly wings. Soft Matter 3:178–182. doi:10.1039/B612667G

Frog Skin: A Giant Leap for Engineering Applications

Yunho Yang

Introduction

The Frog Prince written by the Grimm Brothers is one of the most famous fairy tales among children worldwide. Though there have been many variations of the story over the past century and in a variety of languages, numerous modern versions tell a story of a prince transformed into a frog by a curse. This frog prince must receive a kiss and true love from a princess to break the spell and return to human form (Fig. 12.1). However, what makes it so difficult for the frog is his gross frog figure, with wet, sticky skin and a big mouth. Before he receives a kiss from the princess, the frog must survive predators, hostility from humans, and extreme environments, which all end up teaching him survival skills.

In order to survive his transformation from human being into amphibious animal, the prince has to learn several defensive tactics for survival from his frog friends. One frog approaches him and says that he has to keep his skin very sticky, slippery, and moist by absorbing water through his skin. He also informs the prince that he needs to be able to control his body temperature as a frog, or else he will die very soon. "Frogs and toads have a 'Lycra' type skin that protects them from injury and disease. It comes in a rainbow of color and patterns" (Bernstein and Woods 2010) and even allows frogs to breathe underwater. Another interesting skill that the prince learns from one of his friends, the tree frog, is utilizing toe pads to help him climb moisture-rich surfaces. By the end of the story, the prince learns to live life as a frog and acquires many unique skills as a frog that he would never have been able to do as a human. Though *The Frog Prince* is a children's story, frog skin's surface structure, color variations, and stickiness are protective features that have been widely explored by researchers for engineering applications (Fig. 12.2).

Y. Yang (✉)
Mechanical Engineering, McCormick School of Engineering
at Northwestern University, Evanston, IL 60208, USA
e-mail: YunhoYang2013@u.northwestern.edu

M. Lee (ed.), *Remarkable Natural Material Surfaces and Their Engineering Potential,*   135
DOI 10.1007/978-3-319-03125-5_12, © Springer International Publishing Switzerland 2014

Fig. 12.1 The princess and the frog prince (Illustration by Hye Yeon Yang)

Fig. 12.2 Adapting to life as a frog (Illustration by Hye Yeon Yang)

Moist Film on Frog Skin

Respiration Through the Film

Many frogs have a thin film of water formed on their skin, and it is necessary for them to retain it in order to breathe through their skin (Fig. 12.3). Skin has an important role for frogs, as it acts as a respiratory organ. The oxygen first gets dissolved into a thin, moist film, from where it is diffused into the blood capillaries through holes on the skin (Mackean 2004). In this way, frogs can breathe through their skin even when they are swimming in water. However, frogs need to remain in the vicinity of a water source, because their skin can lose water and dry very easily. This does not prevent frogs from venturing off though, because special adaptations help retain critical moisture. For example, some frogs are adapted to secrete a thick mucous that prevents water from escaping and their skin from drying out, which is why they will feel very slimy (Science Score 2012). Such frogs include those of the genus *Notaden*, which secrete yellow mucus on their dorsal skin that also acts as an adhesive, allowing bondage to many different types of materials, whether polar, nonpolar, moist, or cool. The mucus also encourages cell growth, allowing wound healing (Suárez 2011).

Fig. 12.3 Skin functions as a respiratory organ for frogs (Photo courtesy of Science Score (blog. sciencescore.com))

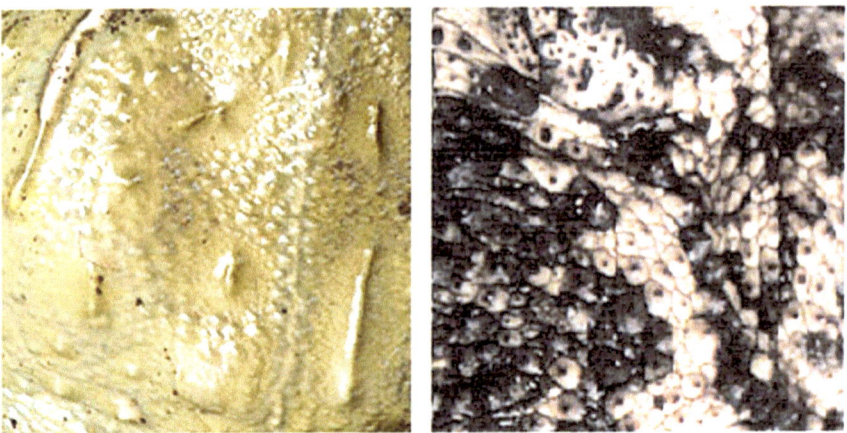

Fig. 12.4 Moist skin of a frog (*left*) vs. dry, warty skin of a toad (*right*) (Photo by Amy Snyder, ©Exploratorium, www.exploratorium.edu)

Lubrication

Another function of the moist film is the lubrication effect it imparts (Fig. 12.4). The pond frog is an example of a moist frog, as it has slippery skin due to the mucous glands on the skin. Clear fluid is secreted from these mucous glands, covering the skin of the frog and moistening it (Els and Henneberg 1990). When the moist film retains certain thickness levels, the frictional effect becomes relatively smaller on the skin than the dry skin. As a result, the pond frog with a continuous fluid film on the skin would be safe from certain minor external physical damages. In addition, such frogs may evade capture from predators due to their slippery skin.

Color Variation

Defensive Strategy

Color variation is one of the powerful defensive tactics of frogs that help them blend in with their natural surroundings. Some frogs such as dart frogs often display their bright skin with specific colored patterns to warn predators that they are highly toxic. The one in Fig. 12.5 usually looks brown and plain, but when it is threatened or in danger, it will expose its belly, which is fire-red. In fact, it is called the fire-bellied toad for this reason (Tesler 1999). Even though not all frogs change their skin colors like chameleons, color variations of frogs are still interesting topics not only for scientists but for the general public as well.

Fig. 12.5 The red belly of *Bombina orientalis* (Photo by Amy Snyder, ©Exploratorium, www.exploratorium.edu)

Source of Color Variation

According to biologists, color variations are generated by frogs from certain pattern changes of chromatophores. Chromatophores are special pigment cells that have an important role in the changing of skin colors and typically consist of three types of layers. The first layer at the bottom is mostly formed with melanophores. Melanophores are dark black or black colored cells, and they contain the melanine pigment that also gives colors to human skin. The color changes in frog skin are mostly controlled by the change of melanophores. The second layer above the bottom-most layer contains iridophores. Iridophores do not have a role in the changing colors but rather the brightness of the skin; they contain purine, which enables them to reflect light to make the frog's skin reflective and silvery like some amphibians and fish. Lastly, ksanthofores are yellowish pigments in the top-most layer. When blue light travels through the top layer, the mixture of blue light and yellowish pigments from ksanthofores makes the frog's skin color look green to human eyes (Yahya 2011). However, as evident in Fig. 12.5, not all frogs have green skin color, which is a result of grains that are distributed differently in the chromatophores that make the color of the skin look different depending on the living environment.

Frog Movement

Toe Pad Surface Structure

There are various types of frogs—some known as torrent or stream frogs can be found on wet rocks near sources of water, while others known as tree frogs can be seen near vegetation like trees and shrubs. However, despite this variation, all frogs have independently evolved to have adhesive pads on their toes. The fact that these pads have developed in different families of frogs is an example of convergent evolution, signifying that the adhesive pad is an optimal evolutionary design for frogs (Barnes 2012).

When we take a close look at the toe pads of some frogs, we can find tribological concepts behind them. Tree frogs' toe pads are good examples of this. Earlier researchers assumed that the mucus would impart a glue-like ability to the toe pads, so that tree frogs could climb stiff or vertical surfaces easily. However, this assumption was found to be wrong, and a consensus was reached with the 'wet adhesion' theory (Federle et al. 2006).

Figure 12.6 shows that the tree frog's toes consist of hexagonal disk-like pads with channel-like spaces separating each pad. "Tree frog toe pads are soft and patterned with a regular hexagonal microstructure of approximately 10 µm diameter epidermal cells separated by approximately 1 µm wide channels; the flattened surface of each cell features a similar but much finer microstructure of approximately 0.1–0.4 µm diameter pegs which originate form hemidesmosomes" (Federle et al. 2006). Based on interference reflection microscopy measurements, it has been concluded that these pegs, also known as nanopillars, directly contact the substrate, producing a formidable amount of shear force (Barnes et al. 2011).

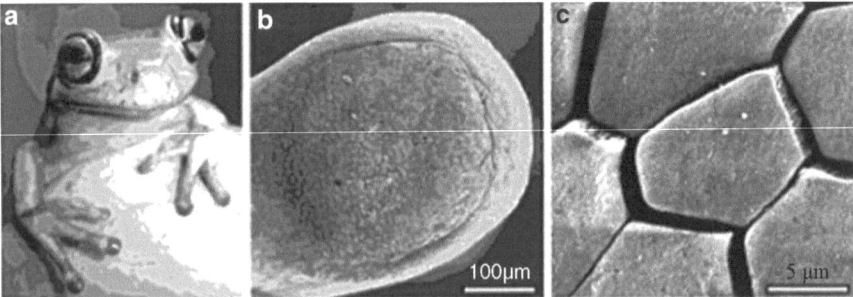

Fig. 12.6 (a) White tree frog, *Litoria caerulea*; (b) scanning electron microscopy (SEM) images of toe pad; (c) SEM images of epidermal epithelial cells in hexagonal shapes (Adapted from Federle et al. (2006), Fig. 1. With kind permission from The Royal Society)

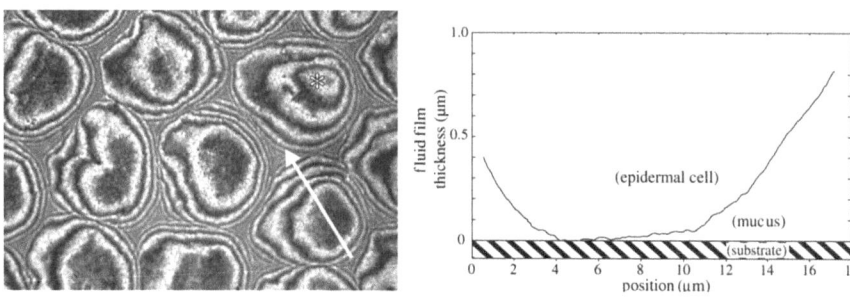

Fig. 12.7 *In vivo* analysis of contact between *Litoria caerulea* toe pad and glass conducted with use of interference reflection microscopy (IRM); wavelength of 436 nm and illuminating numerical aperture (INA) of 0.27, with intensity profile along *white arrow* shown on *left*. Reconstruction of thickness of fluid film on *right*, showing minimum thickness in the center of the epidermal cell (Adapted from Federle et al. (2006), Fig. 2. With kind permission from The Royal Society)

Climbing Mechanism on Moist Surfaces

When a tree frog climbs moisture-rich surfaces, moisture is squeezed out through the channel-like spaces, increasing van der Waals forces between the frog's toes and the surface upon which he is climbing and reducing some lubrication effects. Figure 12.7 shows that the fluid thickness is minimum at the hexagonal toe pad surface, because the fluid is squeezed into the space between each hexagon, causing the toe pad to act like a suction cup (Federle et al. 2006). Further research also reveals that capillary pressure force is a primary mechanism in tensile adhesion, shown by the overall linear relation between toe pad area and adhesion. As toe pad area increases, wet tensile adhesion increases linearly (Lau and Messersmith 2010). Lastly, the grooved channels between the hexagonal cells encourage adhesion by decreasing crack propagation, and pull-off stress is distributed to multiple hexagons so that it is not concentrated on a small contact zone area. Due to these mechanisms, tree frogs, which reside mostly in rainforests, can easily climb up moist surfaces such as wet rocks (Barnes 2012).

Another characteristic of frog toe pads that allows for better adhesion is its extreme softness. All natural materials have a certain surface roughness to some degree, often inhibiting good adhesion due to the inability to come into sufficient close contact. In addition to the nanopillars mentioned in the previous section, which help to increase the area of actual contact, the softness of frog toe pads make them able to conform to the texture and contours of the surface, thereby further increasing actual contact. In fact, frog toe pads were found to be some of the softest biological materials. Measurements of the effective elastic modulus of White's tree frogs, *Litoria caerulea* White, ranged from 4 to 25 kPa, which rival the softness of sea anemone mesogloea and jellyfish jelly, which have an elastic modulus of about 10 kPa. To put these values into perspective, the elastic modulus of tooth enamel is 60,000 MPa (Barnes et al. 2011).

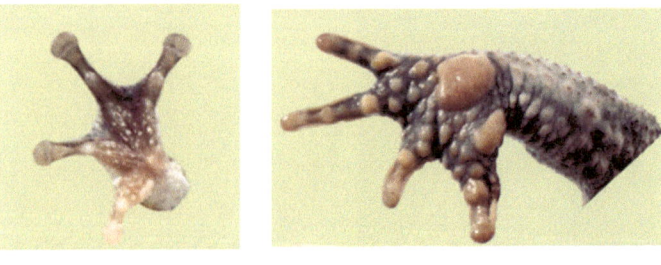

Fig. 12.8 Front foot of tree frog with rounded toes (*top-left*), front foot of toad (*top-right*), and hind food of bullfrog (*bottom*) (Photo by Amy Snyder, ©Exploratorium, www.exploratorium.edu)

Variations of Toes Among Species

A typical frog has two front legs with four toes each, whereas the two back legs have five toes each. Like the example of the tree frog described earlier, frogs that live on the ground have the ability to walk and climb on wet and smooth surfaces by using suction cup-like pads. On the other hand, some aquatic frogs have longer, stronger legs with webbed back feet in order to swim. The webbed hind feet of a bullfrog, as shown in Fig. 12.8, increase water resistance as the long legs engage in paddling motions due to their larger surface areas compared to those of land frogs (Tesler 1999). This advantage enables frogs to swim faster when they hunt for prey in water or escape from predators.

Engineering Applications

The Frog Prince's Secret

To resume discussing the story about the frog prince in the beginning of the chapter, after the frog transforms into a prince by receiving a true kiss from the princess, the prince returns to life as a human being, as in Fig. 12.9. However, to twist the story a little bit… Couldn't it be possible that the prince misses life as a frog at times? After all, he is no longer able to swim and breathe under water, climb up trees, and change colors. Though he gains a beautiful wife and dispels the magic that bound him to life as a frog, he also loses the amazing abilities of being a frog.

Likewise, some mechanisms and functions that frogs have for survival have amazing features that human beings cannot directly follow or mimic. As intelligent beings, however, we can still research and study frogs in order to develop real-world applications.

Grip, Anti-slip, and More

Advances in biomimetics have been made in which toe pad replicas have been produced. At the University of Glasgow, an early attempt at creating a replica featured polydimethylsiloxane (PDMS) with a hexagonal pattern mimicking toe pads' epithelial cells. Later, the Max Planck Institute for Polymer Research in Mainz,

Fig. 12.9 A twist to the frog prince story (Illustration by Hye Yeon Yang)

Germany created a replica that had hierarchical micro- and nano-structures of epithelial cells using hydrophilic PDMS. The replication of such patterns can be applied to adhesive tapes in order to make them stronger. Because detachment of adhesive tapes is achieved when a crack propagates from the point at which it is peeled, peeling is easy if the energy goes into one crack. However, introducing patterns such as the hexagonal one seen on frog toe pads causes the energy to be distributed among all the grooves, ultimately necessitating more force to fully peel off. In fact, micropatterned adhesives can require up to 3 times more force to peel. Furthermore, filling the micropatterned grooves with air or oil, much like the mucus- and fluid-filled channels on frog toe pads, has shown an increase in adhesion of up to 30 times (Barnes 2012).

For example, the wet adhesion function of the tree frog's toe pads can be applied to the grip of car tires on wet roads and any kind of anti-slip or holding device. In addition, unique colors and patterns of frogs have already been adapted widely for military utilizations in order for soldiers and vehicles to be more camouflaged in nature during battle. Finally, swimming fins mimic the shapes of webbed frog feet for faster and more dynamic swimming. There are still many other interesting features of frogs that have not been previously introduced, such as toxic tactics, super-flexible skin, frog locomotion… and perhaps even happy life in the castle with a beautiful princess.

References

Barnes WJP (2012) Adhesion in wet environments: frogs. In: Bhushan B (ed) Encyclopedia of nanotechnology. Springer, Netherlands, pp 70–83. doi:10.1007/978-90-481-9751-4_257

Barnes WJP, Goodwyn PJP, Nokhbatolfoghahai M, Gorb SN (2011) Elastic modulus of tree frog adhesive toe pads. J Comp Physiol A 197(10):969–978. doi:10.1007/s00359-011-0658-1

Bernstein M, Woods M (2010) Frog skin provide 'kiss of death' for antibiotic-resistant germs. http://www.acs.org/content/acs/en/pressroom/newsreleases/2010/august/frog-skin-may-provide-kiss-of-death-for-antibiotic-resistant-germs.html. Accessed 16 Nov 2012

Els WJ, Henneberg R (1990) Histological features and histochemistry of the mucous glands in ventral skin of the frog (Rana fuscigula). Histol Histopathol 5(3):343–348

Federle W, Barnes WJP, Baumgartner W, Drechsler P, Smith JM (2006) Wet but not slippery: boundary friction in tree frog adhesive toe pads. J R Soc Interface 3(10):689–697. doi:10.1098/rsif.2006.0135

Lau KHA, Messersmith PB (2010) Wet performance of biomimetic fibrillar adhesives. In: von Byern J, Grunwald I (eds) Biological adhesive systems. Springer, Vienna, pp 285–294. doi:10.1007/978-3-7091-0286-2_19

Mackean DG (2004) Frogs—an introduction. http://www.biology-resources.com/frog.html. Accessed 18 Nov 2012

Science Score (2012) Did you know that frogs breathe through their skin? http://blog.sciencescore.com/fun-facts-for-kids/did-you-know-that-frogs-breathe-through-their-skin. Accessed 18 Nov 2012

Suárez JC (2011) Bioadhesives. In: da Silva LFM, Öchsner A, Adams RD (eds) Handbook of adhesion technology. Springer, Berlin/Heidelberg, pp 1385–1408. doi:10.1007/978-3-642-01169-6_53

Tesler P (1999) The amazing adaptable frog. http://www.exploratorium.edu/frogs/mainstory/index.html. Accessed 18 Nov 2012

Yahya H (2011) The pigment cell that gives color to frog skin: chromatophores. http://harunyahya.com/en/works/42074/the-pigment-cell-that-gives. Accessed 19 Nov 2012

Michelle Lee

Spiderman's Web: Not So Fictional After All?

The sound of people screaming below and cars honking floods the woman's ear as she slowly steps backward, eyes wide with fright (Fig. 13.1). She can smell the terrible, toxic breath from the slimy, reptile monster looming in front of her as it approaches her with a sharp-toothed grin. She spots a co-worker with blood running down his cheek hiding and trembling behind an overturned desk, but she knows that he won't step out from his hiding place anytime soon to help her. Maniacal laughter erupts from the cavernous mouth of the monster, startling her, and she loses her balance on a coffee mug, sending her toppling backwards into the windy air, 25 floors above ground. The sight of her co-worker's horrified face is the last thing she sees before windows and lights rush past her as she falls to her dea—

Her heart stops. Not because she's dead, but because she can't believe it: she is hovering less than a yard off the ground, enveloped in a sticky, bouncy… *web*. People are shouting and footsteps approach her from all sides as she squirms and screams. When she looks at everyone's faces, she notices that they are all looking up towards the sky with one word on their lips. She follows their gaze to see a red and blue figure swinging away on a cable as everyone whispers with awe, "Spiderman!".

It is stories like these that have enabled Spiderman to win a special place among the hearts of many Americans, young and old. Whether it is in the original comic series or at the theaters for the showing of the newest Spiderman movie, people adore the action, near-death experiences, and life-saving accomplishments of this hero. However, no matter how loved Spiderman is, it is well known that he and his amazing web are just components of a fictional story.

Particular gratitude to Rose Gruenhagen for allowing me to consult her draft on spider silk.

M. Lee (✉)
Mechanical Engineering, McCormick School of Engineering
at Northwestern University, Evanston, IL 60208, USA
e-mail: MichelleLee2013@u.northwestern.edu

Fig. 13.1 A spider on its web

...Or are they? Perhaps this chapter will change your mind on how real or fictional the abilities of his web are, in addition to revealing several other remarkable properties of spider webs that truly deserve nationwide attention.

In fact, spider silk has begun to garner public and media attention recently due to a display at the Victoria and Albert Museum in London. During the year 2012, an exhibit displayed a stunning piece of clothing: golden yellow in hue, adorned with floral embroidery, and with cascading tassels down the front. It wasn't necessarily the style or the design—although it *was* remarkably beautiful—that landed it its own exhibition in one of the world's finest museums. Instead, it was because of its maker: golden orb spiders.

Harvesting and using spider silk for fabric is a long lost procedure dating back to its first emergence in 1709, when Frenchman Francois-Xavier Bon de Saint Hilaire made clothing and accessories for King Louis XIV. It has only resurfaced a few times in the past three centuries, but this golden cape was the first serious, successful attempt since (Harding 2012). An 8-year-long endeavor, the creation of this cape took more than one million Madagascar golden orb spiders. The brilliant yellow of the cape—the natural color of the golden orb's silk—is just one of the remarkable properties of this silk. Typical spider silk is so flexible that it can stretch up to 50 % of its length without snapping, and it is four times as tough as Kevlar and 20 times stronger than steel, weight for weight (Purcell 2012). Characteristics such as these make Spiderman's web, which he used to swing between buildings, catch victims, and perform many other stunts, seem to be closer to reality than originally thought.

In this chapter, we will take a look at the structural and surface properties of various types of spider silk that enable such remarkable performance as a fiber, as well as its past and potential uses in engineering.

One Web, Many Types of Threads

Though there are various types of spider webs such as tangle and sheet webs, the orb web (Fig. 13.2) is the most common and is spun and constructed by orb-weaving spiders, *Araneus* and *Nephila*. Among orb webs, there are variations in design between species, the individual spider, or even on a day-to-day basis. For example, some webs may be two-dimensional and others may be three-dimensional, and some—usually those spun by young spiders—may be more circular than others. In addition, if the spider is hungry, the web will most likely be smaller.

Orb-weaving spiders are capable of producing seven different types of silk threads, each with distinct combinations of amino acid composition, function, and gland and spinneret used in production. For example, minor silk is produced in a gland called the minor ampullate, is spinned by the anterior/median spinneret, and is used in construction of the orb web frame. On the other hand, cocoon silk is produced in the cylindrical gland, is spinned by the median/posterior spinneret, and is primarily used for reproductive purposes. Wrapping silk is produced in the aciniform gland, is spinned by the median/posterior spinneret, and is utilized to wrap captured prey. There are several other silks whose functions range from attachment, prey capture, and construction of the radial parts of the orb web (Saravanan 2006). This chapter will focus on dragline silk and capture threads.

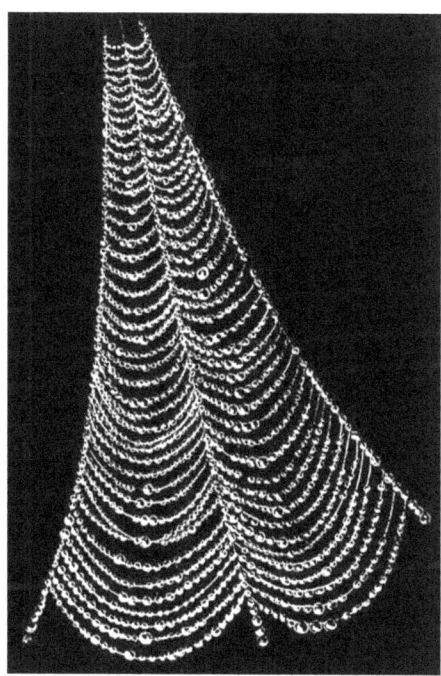

Fig. 13.2 A sector of an orb web laden with dew in an 1890 publication of *Popular Science Monthly* (Reprinted from McCook (1890))

Dragline Silk

What Is Dragline Silk?

The structural makeup of spider silk makes it a renowned fiber that was used by mankind in ancient times, long before it took the spotlight in the material sciences research community in the 1950s (Heim et al. 2009). Ancient Grecians used cobwebs to seal bloody wounds, while ancient Australasians used spider silk for fishing. In fact, they still do. Up to World War II, machines with optical devices such as guns, microscopes, and bomb-guiding systems used spider silk as crosshairs, because it is more than 35 times finer than human hair. As if this versatile array of applications is not impressive enough, it turns out that a rope of spider silk not much thicker in diameter than a garden hose could support an Airbus plane—fully loaded (Gerritsen 2002). How is it possible that spider silk—the same light, sticky silk that could be wiped away with a single flick of a finger—is capable of all this? The answer lies in the structural properties of a specific spider silk known as dragline silk.

Dragline silk, known to be the toughest of all types of silk, serves as the framework of the spider's web, such as the one shown in Fig. 13.1, as well as its lifeline (Heim et al. 2009). It is also coated in perfume to attract mates and is used to store food and eggs (Saravanan 2006). Despite inevitable variations in the material properties of dragline silk across different species of spiders, it is still found to be tougher than most biological fibers and even man-made fibers (Swanson et al. 2007).

Structure of Dragline Silk

Dragline silk is produced in the major ampullate gland and spinned by the anterior/median spinneret. The mean diameter of dragline silk is about 7 μm, but it varies along the length of the fiber ±20 % due to the way the silk is produced. Dragline silk has a skin-core structure, where the skin is weak, and the core is made of twin filaments that are stuck together and have a circular cross section (Saravanan 2006). The silk is made of proteins called fibroins that feature repeated amino acid sequence motifs, forming stiff crystalline structures in a more elastic matrix. It is hypothesized that the variability of these motifs are responsible for the differences in properties among dragline silks from different species. For example, high tensile strength is thought to be a product of beta sheets formed from motifs rich in alanine and glycine. On the other hand, extensibility is a property hypothesized to come from motifs that produce helical, spring-like regions in the silk. In this way, sequence motifs determine what kind of properties that a specific silk will exhibit (Swanson et al. 2006).

Two Types of Capture Threads

The Purpose of Capture Threads

While spiders use dragline silk for web framework and as a lifeline, they also spin threads to capture their prey. Since capture threads' main purpose is to stop prey, they have various properties that make them well suited for this task. Like dragline silk, they are incredibly tough and elastic, since they must absorb the high kinetic energy of an out-of-plane load in maximum deflection (Vollrath 2000). However, strong adhesion of the silk is also crucial for retaining their prey and giving spiders time to locate and reach the captured insect before they can escape. Two kinds of capture silks have been developed over time by spiders to address this need: cribellar thread and viscous thread (Sahni et al. 2011).

Cribellar Thread

Cribellar thread was the first type of capture thread ever to be spun by spiders, but as other tactics for capturing prey emerged among spiders over time, some lost the ability to produce cribellar thread. However, it is still used by about 3,700 species of spiders in 20 different families, helping them retain prey for longer so that they have more time to locate and subdue the doomed insect. Operating independently of glue like substances or viscous fluids, these dry threads consist of a few large, supporting fibrils surrounded by an outer sheath of fine, looping protein fibrils (Opell and Schwend 2009).

Produced from a spinning plate called the cribellum, the fine fibrils are drawn out, or hackled, from thousands of spigots using the calamistrum, a bristly, setal comb on the metatarsus of the spider's fourth leg (Opell and Schwend 2009). This combing charges up the fibrils so that they repel each other, which causes them to puff and resemble a nanoscopic 'wool yarn'. These fine fibrils cling to the pair of thick, support fibers that originate from the main spinnerets, completing the assembly of cribellar thread (Vollrath 2006). Refer to Fig. 13.3a–c for photos of cribellar thread, cribellum spigots, and the cribellum of a female adult Waitkera waitakernisi, respectively. Figure 13.4 shows the cribellum and calamistrum.

Eleven species of the almost 3,700 species that spin cribellar threads are primitive, producing nanofibers that are non-noded and cylindrical. The rest of the cribellate species produce regularly noded nanofibers (Sahni et al. 2011).

Viscous Thread

Viscous capture threads differ from cribellar threads in basic structure. Viscous threads, also called viscid silk, feature a pair of soft, stretchy fibers that are sheathed in aqueous adhesive glue. As opposed to the fibers of cribellar threads, which are

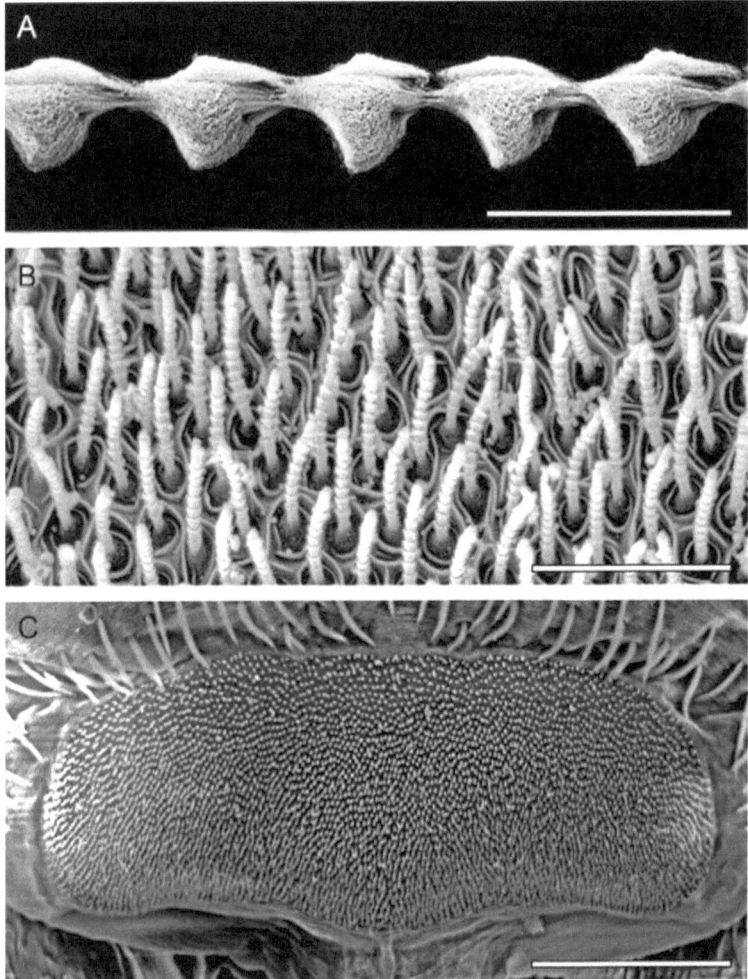

Fig. 13.3 (a) Cribellar thread (scale bar = 200 μm), (**b**) cribellum spigots (scale bar ‑ 10 μm), and (**c**) cribellum (scale bar = 100 μm) of a female adult *Waitkera waitakerensis* (Reprinted from Opell and Schwend (2009). With permission from Elsevier)

produced in single spigots, viscous threads are spun from a triad of spigots and their respective glands: one that produces the axial fiber and two that secrete adhesive glue. Initially, the glue sheaths the fibers evenly, but it soon begins to form droplets that are regularly spaced out. These droplets, which line the thread at about 30 or less droplets per mm, are not as frequent as the nodes of derived cribellar threads, which occur in such small intervals that there are about 170 points per $μm^2$ (Sahni et al. 2011). The glue droplets are 7–29 μm and damp the impact of flying prey by absorbing and dissipating their kinetic energy. The water content of the sheath of glue makes the overall viscous thread more elastic and tough by plasticizing the

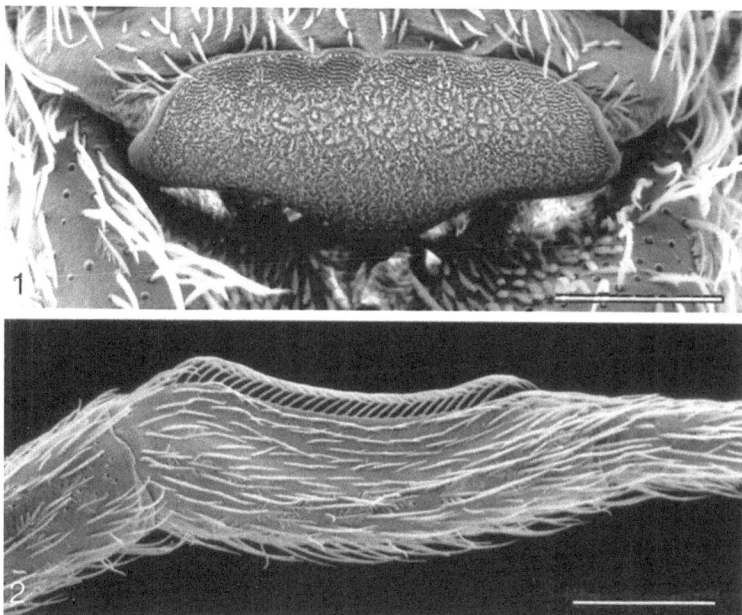

Fig. 13.4 *Top*: Cribellum (scale bar = 150 μm) of a female adult *Miagrammopes animotus*. *Bottom*: Her calamistrum (scale bar 250 μm) (Reproduced with kind permission from Brent D Opell (2001))

core threads (Saravanan 2006). Viscid silk is considered to be more successful for spiders than cribellar threads, as evidenced by the great increase in diversity of *Araneoidea* compared to *Deinopoeidea*, the cribellar-thread producing lineage (Hawthorn 2002).

Adhesive Properties of Capture Threads

Interlocking Mechanism

Of the several adhesive mechanisms that cribellar threads utilize to stick to a variety of surfaces, both the primitive and evolutionarily derived species use an interlocking system much like the Velcro. In this type of adhesion, the looped, woolly sheath of fibrils snags on insects' setae—hairy, bristle-like features—or other surface irregularities (Hawthorn 2002).

Van der Waals and Capillary Forces

Cribellar threads are also able to adhere to extremely smooth surfaces such as glass, polished steel, and graphite, all cases in which the interlocking mechanism is

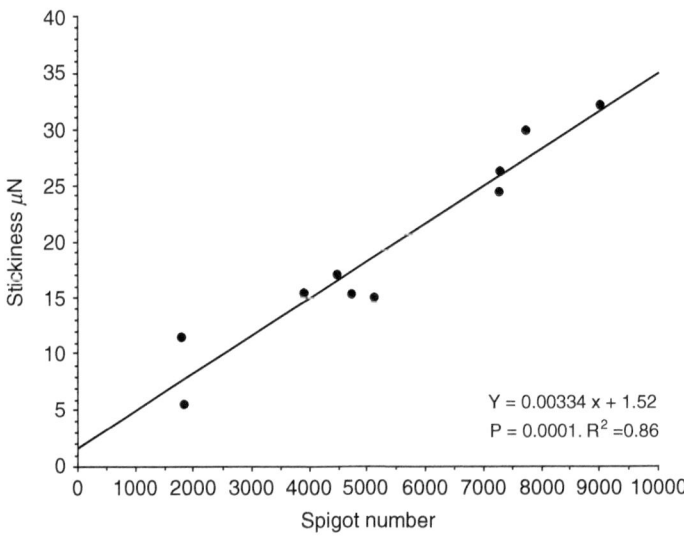

Fig. 13.5 Plot showing the direct correlation of spigot number and stickiness of cribellar thread (Reprinted from Opell and Schwend (2009). With permission from Elsevier)

rendered useless. In fact, cribellar threads hold more securely to the hard, smooth surface of a beetle's forewing than the bristly, dorsal thoracic segment of a fly (Hawthorn 2002). The stickiness of cribellar thread was found by Opell and Schwend to be directly correlated to the number of cribellar spigots; the greater the spigot number, the stickier the thread was (see Fig. 13.5) (Opell and Schwend 2009).

The superb stickiness of cribellar thread that allows it to stick to smooth surfaces is made possible by two adhesion mechanisms, the first being van der Waals forces. Initially observed in 1873 by Johannes Diderik van der Waals, van der Waals forces are weak intermolecular interactions that are present between particles, molecules, or atoms (Mortazavi and Nosonovsky 2012). In an experiment in which non-noded and noded fibers were tested for adhesion in the presence or absence of humidity, researchers found non-noded fibers to have adhesive behaviors that were independent of humidity. Therefore, it was concluded that primitive, or non-noded, cribellar threads use van der Waals forces for adhesion to smooth surfaces (Sahni et al. 2011).

Derived, or noded, fibers had better adhesion at higher values of humidity, indicating that capillary forces were at play (Sahni et al. 2011). Though capillary forces are negligible on the macro-scale level, they are critical at the micro- and nanoscale. Capillary forces exist at the interface between gas, liquid, and solid surfaces, and they cannot be present in a dry environment. Capillary forces can be observed in a humid environment when two solid particles are in contact. Because of the humidity, a meniscus forms from the condensation of water vapor at the contacting interface between the particles, resulting in an attractive force because of the surface tension. Negative Laplace pressure also plays a role in the attraction between the two particles (Mortazavi and Nosonovsky 2012).

Future Engineering Applications

Medical Applications

New biopolymer materials are also a possibility with the manipulation of spider silk proteins. For a spider's purposes, spider silk proteins are mainly employed as a linear thread. However, researchers are attempting to formulate novel materials from newly arranged spider silk proteins, engineering them into various macroscopic shapes and structures such as spheres and capsules. Films can also be made by pouring silk protein solution onto a surface and letting the solvent (e.g., water) evaporate. The spider silk proteins then assemble themselves into a sturdy film as the solvent evaporates. The success in creating such a film opens new doors for scientists to explore direct control of silk protein assembly, allowing the possibility of inventing new biomaterials specifically tailored with remarkable properties unable to be achieved by already existing materials (Römer and Scheibel 2008). Figure 13.6 shows some of the ways in which these biomaterials may be used in the medical field.

Though the sight of a spider and its web may make some people's skin crawl, spider silk—believe it or not—is quite compatible with the human body, including our skin. In the field of plastic reconstructive surgery, novel matrices for skin repair are needed urgently. According to researchers, the ideal biomaterial candidate should not elicit an immune response from the receiver's body. Furthermore, it should promote attachment, growth, and proliferation of skin cells. Finally, when the time comes for degradation as new tissue grows, it should not release substances that may be toxic to the body in any way. Dragline silk fulfills these requirements. Currently, all skin substitutes both naturally derived (e.g., collagen) and man-made (e.g., polylactide-co-glycolide polymers) have dissatisfying mechanical properties *in vivo*, or inside the living body, leaving none available for clinical use as fully functional skin. Dragline silk has been under the research spotlight as of late because of its extensibility, toughness, and stability at large temperature ranges. It is also insoluble in many solvents and is biodegradable. Another critical property that keeps spider silk as a good biomaterial candidate is its biocompatibility. Unlike silkworm silk, which has also been studied for this specific purpose, spider silk does not have an immunogenic sericin coat, therefore eliciting a much less severe immune response (Wendt et al. 2011).

Fredriksson et al. implanted recombinant spider silk (4RepCT) into rats and found that using spider silk as a scaffold for tissue regeneration offers a number of advantages. In addition to being biodegradable, mechanically superior, and biocompatible as mentioned above, recombinant spider silk can be tailored to best promote tissue regrowth by resembling the morphology of the replaced tissue. Finally, because it is produced by genetic engineering in the lab, there is less risk of contaminating hosts with a biohazardous substance. The outlook on using recombinant spider silk is positive, as researchers observed capillaries and cells forming among the 4RepCT fibers merely 1 week after being implanted (Fredriksson et al. 2009).

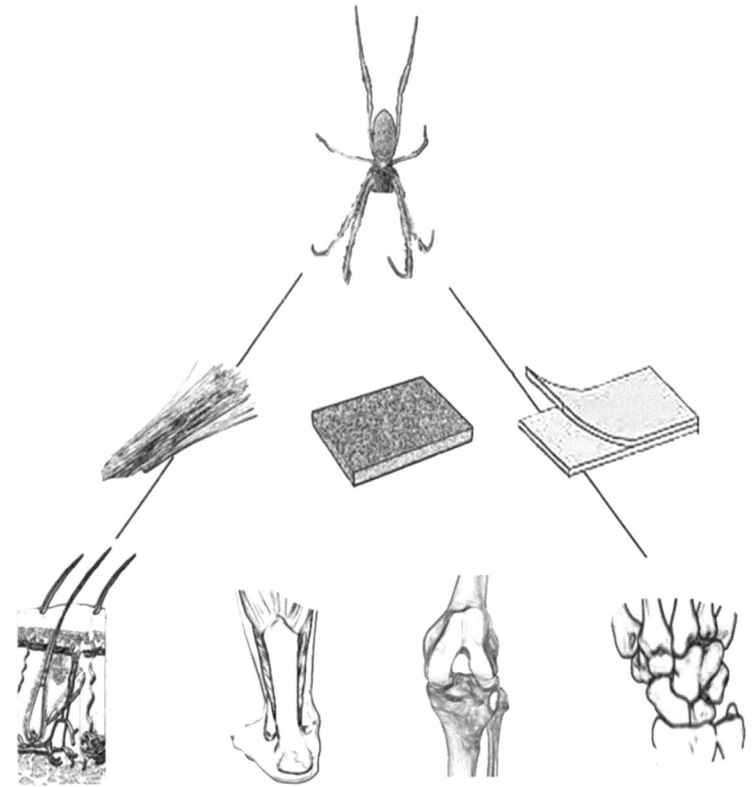

Fig. 13.6 Spider silk, which can be produced in the form of fibers, porous scaffolds (made from solubilized silk protein), and coatings, can be useful in tissue engineering of skin, tendon, cartilage, and bone (Reprinted from Allmeling et al. (2013), Fig. 36.3. With kind permission from Springer Science + Business Media)

In 2004, a bioengineer at Tufts University named David Kaplan also showed that stem cells are more than willing to grow around silk proteins. Theoretically, then, a silk sponge could be used as a scaffold to fix torn muscles or broken bones (Griggs 2012).

The usage of spider silk in peripheral nerve regeneration has been considered as well, and Bruns et al. found immortalized Schwann cells (ISC) to prefer attachment to spider silk, confirming the possibility of its usage as structures for directed nerve guidance. The high elasticity and strength of spider silk is also a plus for successful integration in the target tissue. When the researchers cultivated ISC on spider silk constructs over the course of 2 weeks, they observed the ISC proliferating along the silk strand, as shown in Fig. 13.7 (Bruns et al. 2010).

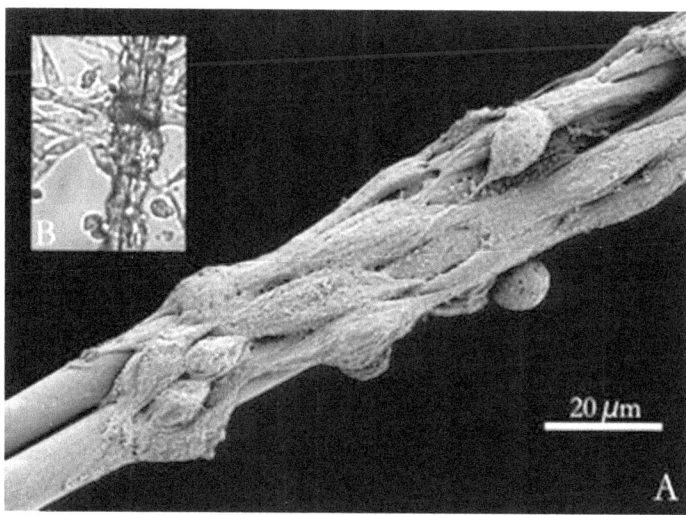

Fig. 13.7 (*A*) Attachment and proliferation of immortal Schwann cells (ISC) on spider silk shown via scanning electron microscopy (SEM) (×1,000). (*B*) Confocal microscopy image (Reprinted from Bruns et al. (2010), Fig. 3. With kind permission from Springer Science + Business Media)

Spider Silk in Military Applications

Though spider silk's properties are covetable, perhaps one of the largest issues with making items out of spider silk is the limited availability of it. Generally, there exist three ways to produce spider silk on a scale large enough for human use: chemically synthesizing the silk, extracting the silk from living spiders, and using recombinant DNA and genetic engineering to produce them. Chemical synthesis is not realistic with the current level of knowledge that chemists have of spider silk. Extracting it from living spiders is very inefficient (Lewis 2001). The golden cape introduced at the beginning of this chapter, which took 8 years and one million spiders to make, is a perfect example of how long the process would take to make just one garment. Furthermore, according to researchers, the territorial and cannibalistic traits of spiders make it unfeasible to raise them in captivity like silkworms. Therefore, scientists aim to genetically transfer the silk-producing ability from spiders to other agents that can produce it in the necessary quantities (U.S. Army 1997). Such agents include bacteria, yeast, tomato plants, crops, and even goats (Discovery 2012).

Perhaps one of the most intriguing uses of spider silk for human purposes is in battle-wear. The use of spider silk in battle is nothing new: Genghis Khan issued silk vests to all his horsemen to protect them from arrows in combat. Now, Dutch researchers are creating bulletproof vests by genetically engineering goats to produce milk that contains spider silk protein. After spinning and weaving the goat milk, the resulting material has ten times the strength of steel.

However, the Dutch researchers are not stopping there. Instead, they are teaming up with the Forensic Genomics Consortium in the Netherlands to create real, human, bulletproof skin. They produced samples of this skin by growing human skin around the bulletproof material from the genetically engineered goats. So far, the resulting bulletproof skin can only stop bullets at reduced speeds, meaning that it does not meet the standard for current bulletproof vests, which is the ability to stop a normal-speed bullet fired from a .22 caliber rifle. However, these scientists hope to one day replace keratin, the protein responsible for making human skin tough, with spider silk protein, creating the ultimate, bulletproof human. Though this seems almost impossible—after all, isn't Superman supposed to be just a comic-book fantasy?—scientists say that it can be done by adding the genes of a spider that produces silk to the genome of a human (Bates 2011).

Spider silk has also recently caught the attention of the United States Navy, who has decided to fund research at Utah State University for 2 years. Based on the observation that spider webs succeed in adhering to surfaces in wet environments such as rainforests and near streams, Navy officials are hoping to know the potential of spider silk as an underwater anchoring adhesive. Martin Lewis, head of spider silk research at Utah State, and his researchers are investigating the feasibility of manufacturing a synthetic replica of the silk material. What is even greater about this endeavor is that the research will be primarily driven by graduate and undergraduate students (Military Times 2013).

Spider silk is a truly impressive natural material with a diverse array of engineering applications, ranging from biomaterials in the medical arena to intriguing uses in battle wear and equipment. Its structure and adhesive properties have been under the spotlight of scientific research for years, and ongoing research may reveal even more discoveries that could very well make a substantial impact on modern day engineering endeavors.

References

Almelling C, Radtke C, Vogt PM (2013) Technical and biomedical uses of nature's strongest fiber: spider silk. In: Nentwig W (ed) Spider ecophysiology. Springer, Berlin Heidelberg. doi:10.1007/978-3-642-33989-9_36

Bates D (2011) 'Making science-fiction a reality': bulletproof human skin made from spider silk and goat milk developed by researchers. http://www.dailymail.co.uk/sciencetech/article-2026645/Bulletproof-human-skin-spider-silk-goat-milk-developed-scientists.html. Accessed 8 Aug 2013

Bruns S, Stark Y, Marten D, Allmeling C, Kasper C, Stahl F, Scheper T (2010) A preliminary study on spider silk as biomaterial for peripheral nerve regeneration. ESACT Proceedings. Cells Culture 4:573–578. doi:10.1007/978-90-481-3419-9_99

Discovery (2012) Body armor made from spider silk. http://news.discovery.com/tech/gear-and-gadgets/body-armor-spider-silk-1210152.htm. Accessed 8 Aug 2013

Fredriksson C, Hedhammar M, Feinstein R, Nordling K, Kratz G, Johansson J, Huss F, Rising A (2009) Tissue response to subcutaneously implanted recombinant spider silk: an in vivo study. Materials 2:1908–1922. doi:10.3390/ma2041908

Gerritsen VB (2002) The tiptoe of an airbus. http://web.expasy.org/spotlight/back_issues/024/. Accessed 8 Aug 2013

Griggs J (2012) On the silk road. New Sci 213(2850):36–39

Harding E (2012) The gown of gold spun by a million spiders that revives a lost tradition. http://www.dailymail.co.uk/femail/article-2090608/Reviving-lost-tradition-Cape-silk-million-spiders-unveiled-new-exhibition.html. Accessed 1 Nov 2012

Hawthorn AC (2002) Evolution of adhesive mechanisms in cribellar spider prey capture thread: evidence for van der Waals and hygroscopic forces. Biol J Linn Soc 77(1):1–8. doi:10.1046/j.1095-8312.2002.00099.x

Heim M, Keerl D, Scheibel T (2009) Spider silk: from soluble protein to extraordinary fiber. Angew Chem Int Ed 48:3584–3596. doi:10.1002/anie.200803341

Lewis R (2001) Unraveling the weave of spider silk. http://www.mhhe.com/biosci/genbio/life/articles/article1.mhtml. Accessed 8 Aug 2013

McCook HC (1890) The strength of spiders and spider-webs. Popular Science Monthly 37:42

Military Times (2013) Navy bets on spider silk research with USU funding. http://www.militarytimes.com/article/20130807/NEWS04/308070038/Navy-bets-spider-silk-research-USU-funding. Accessed 8 Aug 2013

Mortazavi M, Nosonovsky M (2012) Polymer adhesion and biomimetic surfaces for green tribology. In: Nosonovsky M, Bhushan B (eds) Green tribology. Springer, Berlin Heidelberg, pp 173–219. doi:10.1007/978-3-642-23681-5_8

Opell BD (2001) Cribellum and calamistrum ontogeny in the spider family Uloboridae: linking functionally related but separate silk spinning features. J Arachnol 29(2):20–26

Opell BD, Schwend HS (2009) Adhesive efficiency of spider prey capture threads. Zoology 112(1):16–26. doi:10.1016/j.zool.2008.04.002

Purcell A (2012) The science of the golden spider-silk cape. http://www.newscientist.com/blogs/shortsharpscience/2012/01/the-science-of-the-spider-silk.html. Accessed 1 Nov 2012

Römer L, Scheibel T (2008) The elaborate structure of spider silk. Prion 2(4):154–161

Sahni V, Blackledge TA, Dhinojwala A (2011) A review on spider silk adhesion. J Adhes 87:595–614. doi:10.1080/00218464.2011.583588

Saravanan D (2006) Spider silk—structure, properties, and spinning. J Textile Apparel Technol Manag 5(1):1–20

Swanson BO, Blackledge TA, Beltran J, Hayashi CY (2006) Variation in the material properties of spider dragline silk across species. Appl Phys A 82(2):213–218. doi:10.1007/s00339-005-3427-6

Swanson BO, Blackledge TA, Summers AP, Hayashi CY (2007) Spider dragline silk: correlated and mosaic evolution in high-performance biological materials. Evolution 60(12):2539–2551. doi:10.1111/j.0014-3820.2006.tb01888.x

U.S. Army (1997) Untangling the web. http://www.ssc.army.mil/about/pao/pubs/warrior/97/nov/silk.htm. Accessed 8 Aug 2013

Vollrath F (2000) Strength and structure of spiders' silks. Rev Mol Biotechnol 74(2):67–83. doi:10.1016/S1389-0352(00)00006-4

Vollrath F (2006) Spider silk: thousands of nano-filaments and dollops of sticky glue. Curr Biol 16(21):R925–R927. doi:10.1016/j.cub.2006.09.050

Wendt H, Hillmer A, Reimers K, Kuhbier JW, Schäfer-Nolte F, Allmeling C, Kasper C, Vogt PM (2011) Artificial skin—culturing of different skin cell lines for generating an artificial skin substitute on cross-weaved spider silk fibres. PLoS One 6(7):e21833. doi:10.1371/journal.pone.0021833

Index

A

Abbott-Firestone Load Curve (AFLC), 109, 110
Adhesion, 18, 59–60, 115–117, 119–125, 130, 140, 141, 144, 149, 151, 152
Adhesive force, 59, 106, 109, 117, 119–122
Adhesive pads, 140
Advancing contact angle, 56, 93
AFLC. *See* Abbott-Firestone Load Curve (AFLC)
Amontons' law of friction, 109
Anisotropy/anisotropic, 94–96, 104–108, 120, 121
Antenna coatings, 93
Anti-adhesive, 84
Antibiotic-resistant, 21
Antibiotics, 20, 21, 23, 25
Anti-fogging, 84
Antifouling, 25
Antireflection coatings (ARCs), 84, 85
Antireflective (AR), 82–86, 88, 121
AR. *See* Antireflective (AR)
Aragonite, 30–34, 38
ARCs. *See* Antireflection coatings (ARCs)
Armor system, 31
Artificial adhesive, 124
Asexual reproduction, 47
Aspect ratio, 121
Autonomic-healing, 2, 3, 11
Autonomic-repairing, 2

B

Bacteria, 15–25, 47, 61, 155
Bending, 8, 65, 67, 70–72
Benthic diatom, 43
Binary fission, 47
Biocide, 19, 20, 25, 61
Biocompatibility, 39, 49, 53, 122
Biocompatible, 122, 153

Biodeterioration, 61
Biofilms, 18–21
Biofouling, 19, 22, 25
Bioinspiration, 3, 91, 125
Biomaterials, 153, 156
Biomimicry, 21, 22, 26, 79, 84, 125
Biopolymer, 38, 153
Biosensor, 49
Bleeding, 3–10
Blood clotting, 3–4
Bone implant, 39–40
Boundary layer, 75, 112
Brick-and-mortar, 32–33
Brittle, 2, 31, 34, 38, 71
Bulletproof, 155, 156
Butterfly, 76, 79, 127–134

C

Calamistrum, 149, 151
Calcite, 31
Calcium carbonate crystals, 31
Capillary forces, 98, 116, 117, 151–152
Capillary network, 9
Capture threads, 147, 149–152
Carbon nanotube (CNT)-based adhesive, 121, 123–125
Cassie-Baxter (CB) model, 57–58
Cassie State, 130
Catalyst, 4–6
Catastrophic failure, 2, 36
Catheter-associated urinary tract infections (CAUTI), 23
Centrales, 46, 57
Ceramic, 31
Chitin, 67, 71–74, 83
Chromatophores, 139
Circulatory system, 9
Cohesive force, 59
Color variation, 135, 138–144

Columnar nacre, 33
Composite, 4–9, 12, 17, 31, 34, 38–40,
    72, 74, 95
Compound eyes, 81
Concrete, 6, 61, 93, 97
Contact angle, 55–57, 68, 93–95, 130, 132
Contact area, 56–60, 109, 112, 121, 124, 131
Contact-splitting principle, 120
Corneal lens, 82
Corneal nipples, 82–84, 87, 88
Corrosion, 5, 10–12, 19, 97
Corrosion protection, 10–12
Corrugation, 69–70
Crack, 1, 4–6, 31, 32, 35, 36, 141, 144
Cribellar thread, 149–152
Cribellum, 46, 149–151
Crystalline cone, 81, 82
Cylinder, 112, 113

D

Damage management, 2
Damage prevention, 2
Dentine, 17
Dermal denticles, 17, 18, 21–22
Dermis, 17, 105
Diatom, 41–51
Diatomaceous earth, 48
Diatom shell, 41, 49
Digital pad, 118
DNA purification, 48
Dorsal, 67, 68, 84, 107, 111, 128, 137, 152
Drag, 67, 68, 84, 107, 111, 128, 137, 152
Dragline silk, 147–149, 153
Dragonfly, 65–76
Drug delivery, 49
Drug delivery vehicle, 49
Dry adhesion, 49
Ductile, 31, 34, 38, 71

E

Elastic modulus, 141
Encapsulation, 4–6, 11, 97
Energy, 6, 23, 26, 31–33, 38, 39, 43, 49, 76,
    86–88, 100, 112, 121, 124, 131, 144, 150
Energy consumption, 112
Epicingulum, 45
Epicuticular wax, 58, 94
Epidermis, 17, 105
Epitheca, 45, 47, 48
Epithelial cells, 140, 143, 144
Epivalve, 45
Equilibrium contact angle, 56, 57, 93

Escherichia coli, 23, 25
Extensibility, 148, 153
Extracellular polymeric substances (EPS),
    18, 19, 83, 131, 145, 153

F

Facet lenses, 81
FEA. *See* Finite element analysis (FEA)
Fibrils, 73–74, 106–110, 149, 151
Fibrin, 3, 4
Fibroins, 148
Finite element analysis (FEA), 72, 73
Finite element model, 6, 37–39
Flexural rigidity, 6, 8, 70
Fluorescence, 49, 130
Fouling, 18, 19, 21, 22, 25
Friction, 22, 39, 100, 104–113, 116, 120, 138
Frictional anisotropy, 104–108
Friction drag, 22
Frog skin, 135–144
Frustules, 43, 45–48, 50
Fuel cell, 99

G

Gecko adhesion, 115–117, 124
Geckos, 115–121, 124
Girdle bands, 45
Glass, 6–8, 12, 31, 41–51, 57, 84, 85, 96–97,
    119, 132, 133, 141, 151
Grooves, 21, 68, 85, 112, 113, 131, 141, 144
Grubbs' catalyst, 5

H

Hackled, 149
Healing agent, 3–6, 8–10
Hertzian theory of elastic mechanical
    contact, 109
Hexagonal array, 82
HGF. *See* Hollow glass fibres (HGF)
Hierarchical structure, 31–33, 46, 58–59, 71,
    76, 94, 95, 117, 118, 121, 122, 125
Hollow glass fibres (HGF), 6–8
Hollowness ratio, 8
Humidity, 116, 117, 152
Hydrophilic, 56, 57, 93, 116, 144
Hydrophobic, 56, 57, 62, 91, 93, 97–99, 116,
    132, 134
Hypocingulum, 45
Hypotheca, 45, 47, 48
Hypovalve, 45
Hysteresis, 93

I

Immortalized Schwann cells (ISC), 154, 155
Implants, 20, 39–40
Index of refraction, 83
Inelastic deformation, 35
Inelastic dissipation, 31
Interfacial tension, 18, 56
Interference lithography, 85
Interlocking mechanism, 37, 151
Internal combustion engines (ICEs), 112, 113
Ion etching, 97
Iridescence, 79, 127, 129–131
Iridophores, 139
ISC. *See* Immortalized Schwann cells (ISC)
Isotropic, 94–96, 100, 121

K

Ksanthofores, 139

L

Lamellae, 117–118, 120, 128, 129, 131
Lift, 67, 69, 70, 75, 127, 133
Light, 49, 57, 69, 79–88, 107, 109, 124, 128, 130, 132, 139, 148
Lipid layer, 124
Load bearing capacity, 73, 109
Lotus, 53–62, 91, 93–95, 97, 100, 130
Lotus effect, 54, 61, 62
Lotus leaves, 53–62, 91, 94, 95, 97, 100, 130
Lubricant, 98, 112
Lubrication, 113, 138, 141

M

Mantle, 30, 45, 47
MAVs. *See* Micro-air-vehicles (MAVs)
Maximum sustainable force, 124
Melanine, 139
Melanophores, 139
Membrane, 19, 67–69, 71, 73, 99, 107, 122, 123
Micro-air-vehicles (MAVs), 76
Microcapsules, 4–6
Microfibres, 9
Microfibrils, 106–109
Micropits, 110–112
Mixed fracture model, 71
Molding lithography, 132
Mollusca, 29, 30, 33, 35, 40
Mother-of-pearl, 29–40
Moth eyes, 79–88
Mucilage, 47
Mucous, 137, 138

N

Nacre, 31–35, 37–40, 117
Nacreous shells, 30–32
Nano-asperities, 38
Nanopillars, 140, 141
Nano-tips, 128, 130, 131
Noded, 149, 152
Non-noded, 149, 152
Nosocomial infections, 25, 26

O

Oberhautchen, 105
Ommatidia, 82
Optical superposition, 81
Optical system, 81
Optics, 4, 7, 35, 49, 76, 81, 83, 84, 86, 88, 130, 132, 134, 148
Orb web, 147
Oxygen, 9, 48, 137

P

Papillae, 58–60, 94–96
Peel mechanism, 120, 125
Pegs, 140
Pennales, 46, 47
Permeability, 111, 112
Photic zone, 43
Photolithography, 85
Photosynthesis, 85
Photovoltaic (PV) systems, 86
Pigment cells, 139
Pigments, 43, 130, 139
Pinning, 131, 149, 155, 157
Piston, 112, 113
Placoid scales, 17
Planktonic bacteria, 20
Plasma-chemical roughening, 97
Plateau honed surfaces, 112
Polydimethylsiloxane (PDMS), 22–24, 132, 133, 143, 144
Polymer-based adhesive, 121, 122
Polymerization, 4–6
Pores, 18, 46, 47, 49, 99, 111, 112, 122
Pressure drag, 22, 75
Prosthetic adhesive, 122
Protein, 3, 18, 32, 33, 37, 38, 48, 49, 67, 71–74, 76, 105, 148, 149, 153–156
Protuberance, 82–84, 87, 88, 98, 99
Purine, 139

Q

Quadrate scales, 128

R

Radiation, 57, 88
Raphe system, 47
Receding contact angle, 56, 93
Recombinant DNA, 155
Reflectance, 84, 85, 128
Relative rigidity ratio, 8
Retinula cells, 82
Reversible adhesion, 18, 115
Reynolds number, 69, 70
Rhabdom, 81, 82
Rhizomes, 53
Riblets, 17
Rice leaves, 91, 93–96, 130
Ridges, 21, 83, 95, 109, 110, 118
Ridging stripes, 128
Ring-opening metathesis polymerization
    (ROMP), 5
Ripple morphologies, 75
RMS. *See* Root mean square (RMS)
Rolling state, 130
Root mean square (RMS), 35
Roughness, 35, 56–58, 95, 106, 116, 141

S

Sandwich structure, 71–72
Scales, 17, 22, 31, 46, 103–113, 124, 125,
    128–130, 132, 133
Scattering, 129, 130
Scintillator, 88
Self-cleaning, 54, 61, 84, 96–97
self-healing, 1–12
Self-repairing, 1, 2, 12
Sessile bacteria, 18
Setae, 116, 118–120, 124, 151
Sharklet AF™, 22–26
Shark skin, 15–26
Shark skin effect, 22
Sheet nacre, 33
Silica, 43–49, 132
Silicone elastomer, 22, 23
Silk, 145–156
Silk protein, 153–156
Sliding, 33–34, 38, 93, 95, 104, 106, 109,
    112, 113, 119, 120
Sliding angle, 93
Slime, 18, 25
Smart adhesion, 115, 116
Smart materials, 2
Snake skin, 103–113
Solar cells, 49, 86, 88, 97
Solar energy, 49, 86–87
Sol-gel, 132

Spatulae, 118–120, 122, 124
Spider, 79, 145–156
Spider silk, 145–156
Spigot, 152
Staphylococcus aureus, 22, 25
Static coefficient of friction, 106
Stomata, 58, 59
Strain, 34, 37–39, 72, 73
Stress-strain curve, 34
Superhydrophilic, 56, 57, 93
Superhydrophobic, 54–58, 61, 68, 84,
    93–100, 130
Superhydrophobicity, 53–55, 57, 58, 62, 68,
    91–94, 130–131
Surface roughness, 56–60, 141
Synthetic adhesive, 124

T

Tablet, 33–39
Three-phase contact line (TCL), 130, 131
Toe pads, 116, 117, 125, 135, 140, 141,
    143, 144
Torsion, 73
Toxic, 23, 25, 26, 138, 144, 145, 153
Transition metal catalyst, 5
Transmission, 35, 84, 85, 98, 112
Tributyltin (TBT), 25
Tubule, 58
Turbulent flow, 22

U

Urinary catheters, 20, 23
UV fluorescent dye, 6, 7

V

van der Waals forces, 18, 116–117, 124,
    141, 151–152
Vascular networks, 1–12
Vein, 71–76
Vein-joints, 75–76
Ventral, 67, 68, 84, 104–107, 109, 128
Ventral scales, 104–107, 109
Viscid silk, 149, 151
Viscous thread, 149–150
Von Mises stress, 72

W

Water repellent, 61, 93
Wavelength, 49, 83–85, 141
Waviness, 34–36, 38, 39

Wax columns, 67, 68
Wear, 2, 39, 104, 105, 108–113, 124, 155, 156
Web, 1, 142, 144–149, 153, 156
Wettability, 55–57, 112
Wetting, 55–57, 59, 127
Wings, 62, 65–76, 80, 83, 84, 127–134

X
X-ray, 69, 87–88

Y
Young's law, 56